C000174693

Science for Heretics

Barrie Condon

Copyright © 2019 Barrie Condon

No portion of this book may be reproduced by any means, mechanical, electronic, or otherwise, without first obtaining the permission of the copyright holder

The moral right of Barrie Condon has been asserted by him in accordance with the Copyright, Designs & Patents Act of 1988.

ISBN: 978-1-9164572-1-8

This edition published by Sparsile Books Ltd, PO Box 2861, Glasgow, UK, G61 9ED.

SPARSILE BOOKS

Contents

Introduction

Heresy, defined in a wide sense, meant a deviation from actions, attitudes and ideas that guarantee a well-rounded and sanctified life.

Paul Feyerabend [1]

This book examines many heretical ideas that strike at the heart of science as we know it. Having myself been a professional scientist for over thirty years this book is therefore essentially a very long suicide note. Inevitably I'm going to offend colleagues and friends as well as complete strangers.

Heretics almost never do well. Those who believe in or even just communicate heretical ideas nearly always pay a price. Science, like religion, has a history of killing its heretics though that doesn't usually happen nowadays; vilification, ostracism and the shredding of reputation are now its punishments.

Indeed, some view science as increasingly like a religion in the sense that it too has many followers who believe implicitly in its world view. These adherents see it as both the sole truth and the only way forward for mankind.

On the other hand, comparing science to religion is too simple an analogy: science is in a far more favoured position than any religion. Let's take an example.

Suppose that one day the Pope appears on the balcony of the Castel Gandolfo Apostolic Palace to make an announcement to the whole world. In full papal pomp and splendour, and in front of myriad TV cameras, he clears his throat uneasily and says: "Fellow Christians, you know this chap Jesus we've been telling you about for all this time? Well it turns out there are nineteen more of them."

How does the world react?

Difficult to say exactly but the results would not be pretty.

Yet the funny thing is that only a few years ago respected theoretical physicists did something very similar. As a physicist myself, this was the moment that turned my already healthy scepticism into something much darker.

What these physicists were talking about was nothing less than the whole universe. For most of the twentieth century the firm consensus had been that the universe had started with the Big Bang, that it had exploded into existence out of nothing and that it has been expanding outward ever since. Although other theories had come and gone since this idea was first mooted in 1927, the Big Bang Theory, and the theory of gravity that underpinned it, had become received wisdom: gospel, almost.

This had got to the point where the big question wasn't whether this was true or not, it was about what the ultimate consequences would be. Would the universe expand forever, each star flying further and further away from the others, burning up all its fuel and becoming a dead cinder? In this version the whole universe would end in a dying whimper.

Or alternatively, would there be enough matter in the universe for gravity to overcome the explosive effects of the Big Bang, eventually drawing everything back into one great ball of matter in a process called the Big Crunch, perhaps for the whole cycle of expansion and contraction to begin again?

For decades this debate went back and forth. Research effort was concentrated in trying to work out just how much matter there was in the universe because it was thought this would govern whether the universe faded away or whether it would yo-yo for eternity. For a long time, it looked as if it could go either way because the estimated mass of the universe seemed perilously close to the tipping point.

But then cosmology, which is the study of the origin and fate of the universe, had its Castel Gandolfo moment. Respected scientists announced that as well as the mass and energy that made up the known universe, there was also another nineteen times as much that we hadn't known about.

Oh yes, and that this extra nineteen times as much mass and energy was all invisible and undetectable. They called this stuff Dark Energy and Dark Matter which made them sound sort of spooky and sexy.

Then they went even further and said that this Dark Energy acted in the opposite way to gravity: it pushed matter apart rather than pulling it together. This was quite a stretch as nothing even remotely like that had ever been detected, yet now the universe was supposedly teeming with this Dark Energy.

And the reason for making this incredible announcement wasn't that Dark Energy and Dark matter had been observed but rather because they supposedly had to be there to explain experimental measurements that would otherwise undermine some rather precious theories. Specifically, astronomers had found that the stars, rather than slowing down under the effect of their mutual gravitational attraction, were instead accelerating away from each other.

This would make nonsense of the long-held theories that underpinned what we thought we knew about gravity. If cosmologists were to preserve their faith in these theories then a couple of very large rabbits would have to be pulled out of a very big hat.

In other words, Dark Energy and Dark Matter were fudges of, literally, astronomical proportions.

Overnight the followers of science were expected to believe that 95% of the universe consisted of stuff they'd never heard of before and couldn't see, and that gravity, dogmatically considered to be the controller of the evolution of the universe for hundreds of years, now had only a relatively small part to play in it.

And how did the rest of the world react to this stunning announcement? What was the effect when they realised that physics had been so fabulously in error? Did they burn their textbooks or hang the scientists from street posts? Did they never believe a word the scientists said again?

Not at all. Those who had any interest became intrigued by the science fiction-like ideas of invisible matter and anti-gravity forces. In other words, they focused their attention on the rabbits, rather than the fact they had appeared out of thin air.

Dark Energy and Dark Matter aren't the only big rabbits around. Somehow keeping their faces straight, physicists also invoke a score or more entirely new spatial dimensions and infinite numbers of alternate universes, none of which are detectable. They do this to explain away uncomfortable experimental findings, as we will see later. In order to keep their creaking theories intact they employ fudges so massive that they make Dark Energy and Dark Matter look like the most delicate of fine adjustments.

And it is not just the edifice of physics which is built on such shifting sand. Biology and medicine are also teetering on shaky foundations.

But how else can we deal with the universe except by science? Nowadays it's not like there is an alternative to science in the way that, for example, Hinduism can be seen as an alternative to Christianity.

And this is why science is not simply analogous to religion; science has embedded itself far more pervasively into the bedrock of mankind's belief system than any religion. In a world brimming with uncertainty and occasional catastrophe, the apparent stability offered by science and the scientific method to understand a scary universe can seem our only anchor.

Competing theories of religion and politics bombard us and come and go like fashions. How comforting therefore that when an object drops, say an apple from a tree, we think we know exactly what will happen time after time. Not only that but we have faith it will do so without variation because of the law of gravity, exquisitely defined by experiment and the supporting lattice of theory.

Science therefore appears to provide things that philosophy and religion and politics can't, namely the ability to produce the same results each time and, by using the scientific method, enables us to predict events and effects with high degrees of certainty. Science is the comfort blanket of the age.

My job in this book is to try to snatch this blanket away from you, to leave you with nothing to cling onto, to be cast adrift in the numberless complexities of a relentlessly confounding universe. Indeed, it will be argued that complexity is perhaps the only real truth there is in this universe.

I'm going to examine some of the most fundamental difficulties and problems of many aspects of science, from the physics of the very small to the physics of the very large, as well as revealing the inherent limitations in our understanding of medicine, biology and the social sciences.

I'm a physicist by profession so perhaps you may be surprised that I should even mention the field of social sciences. As a practitioner of physics, the hardest of 'hard' sciences, I had spent my career with a rather patronising attitude to those involved in social sciences. I vaguely believed, and I was far from alone in this, that they were lesser scientists because their work was rarely underpinned by complex mathematics-based theory and seemed devoid of anything like physics' beloved universal 'Laws'.

Now, belatedly, I realise that social scientists are the heroes engaged in a titanic struggle with messy reality. Reality, I will argue, that is something that hard scientists have very little to do with.

Though the very title of this book may perhaps reveal some indication of my own opinions, I'm going to try to maintain some semblance of objectivity by examining the various experiments and theories in this book from the viewpoint of three notional characters: The Believer, The Sceptic and the Heretic. They represent increasing levels of doubt about science.

The Believer is pretty sure we've more or less learnt everything there is to know about science, that all that remains is to fill in a few of the final details. During long periods of apparent scientific stability, when no new significant findings have come in to challenge the prevailing scientific orthodoxy, then someone has always stepped forward and proclaimed that mankind was getting to the point where it knew pretty much everything there was to know about science. Probably the most recent popular example of this can be found in the book The End of Science [2]. It must be said that throughout history such people have always been shown to be wildly wrong.

Indeed, the recent sudden need to resort to cosmic scale phantasms like Dark Energy and Dark Matter is a perfect example of how naive this attitude may be and gives an inkling of just how far we are from knowing everything about the workings of the universe.

And this is where the second level of doubt comes in. It simply acknowledges that science is still evolving. This is the view of the Sceptic and pretty much represents the viewpoint of most scientists today. Theory is accepted until new experimental findings show them to be untrue. Theory is then modified, or even discarded and entirely replaced, to better explain these new facts. Sooner or later, as this optimistic view has it, our theory will converge on the truth and we will understand the universe and reality itself.

Such a sceptical attitude is unquestionably healthy. After all, theories in all branches of science come and go all the time. Hanging onto any current theory and considering it to be the immutable truth is very likely to be folly, as the entire history of science compellingly shows. Nevertheless, it is easy to slide from this scepticism into belief. Theories that are provisional easily become orthodoxies not only in the scientific community but also amongst the general public. For example, take any glossy modern-day TV science show and watch the presenter talk as though things like Black Holes or Dark Energy or Dark Matter really exist rather than being, at best, very early guesses based on rudimentary theory. Such presenters may argue they say such things out of expediency. After all, always inserting qualifying phrases such as 'present theory indicates' before every mention of m-branes or quasars or 27-dimensional superstrings or whatever, will make it more boring for the viewer. The presenters might argue that it's more important to get over the ideas and educate viewers in science rather than be overly pedantic and scare them away. Unfortunately, such expediency serves only to harden the current theories into 'facts'.

So, the Sceptic must always be on his or her guard lest they lazily come to regard existing theories as the truth. Nevertheless, the Sceptic believes that, given sufficient progress, one day they will. As to when that glorious day might come is another issue. This book will try to show, by examining the history of a range of sciences, how often the apple cart of accepted theory has had to be overturned to explain challenging new experimental findings. At the very least it will try to show that we must still be very, very far from really understanding how the universe and everything in it works even if the view of the Sceptic is correct.

But now we come to our third character, the Heretic, where the level of doubt is deeper, darker and much, much scarier. The Heretic argues that fundamental limitations in both the scientific method, and also in the way our brains work, mean that we can never understand the universe. Indeed, the Heretic does not believe that the universe works according to underlying laws. He or she believes that the universe just is. That all the laws and principles and equations we use to describe it are just stories we tell ourselves. When some evidence is presented that shows the story is untrue, we discard it and immediately comfort ourselves by telling another one. The Heretic believes such interpretations of the universe will never be anything more than stories and do not represent any meaningful truth.

The Heretic regards our scientific laws as analogous to the lines of latitude and longitude. These are entirely human notions that we project onto the globe and can be very useful in allowing us to navigate it. However, the Earth is entirely unaffected by whether these notions existed or not. Similarly, we project our scientific laws and principles onto a universe that continues to go about its business without any reference to them at all.

So, this book is about the underlying but rarely acknowledged assumptions and limitations of science. At its worst, when taking the extreme heretical view, it seeks to demonstrate that science is a distorting filter that can never show us the truth behind reality, that all scientific and mathematical theories and models are fictions. Sometimes they are well formulated and make a compelling narrative and may even stand the test of time for a while but, ultimately, they are, and will always be, wrong.

If the Heretic is correct and we do not even begin to understand a universe that is full of vast energies and dangerous phenomena, then some of our experiments may come with entirely unexpected and potentially extinction level risks. In essence we may be wandering unknowingly through a minefield. We'll later examine some of the much smaller sized mines we've already tripped over as a species. We've survived so far, but for how much longer.

Though the book is highly critical of science it must be said that, at present, science is indeed all we have. There are unquestionably regularities in the behaviour of the universe, at least here on planet Earth and at this point in time. These regularities can be exploited and will yield many more technological marvels (engineers, like social scientists, are the heroes of this book) and will give us many apparent insights into life and the universe. The scientific method can undoubtedly have utility but that still doesn't mean it is the truth. There may in fact be better ways of dealing with the universe than science but until we realise the fundamental limitations of the scientific method, then we are very unlikely to look for them.

Many of the heretical ideas described in this book have been around for a long time. Ever since the scientific method was adopted many people have been cautioning about the inherent problems with this approach and have criticised science within its own terms and using its own methods. Some have even paid for this questioning with their lives. Nowadays academics who put forward such views risk only ridicule and the withdrawal of grant support rather than burning at the stake. That, at least, is progress.

Unfortunately, the writings of these heretics tend to be on specialised aspects of science hidden deep within rarefied scientific texts where they can easily be glossed over or ignored and then finally forgotten. One of the purposes of Science for Heretics is to draw together for the first time many of these other unsettling ideas from across the scientific specialities and to present them in a form accessible to the general reader who has not received an extensive scientific education.

We will therefore be covering a wide range of science in this book, from physics to medicine, from psychiatry to the social sciences. The disadvantage of this is that in order to cover such a range and yet keep the book within readable limits, the treatment of each science will admittedly not be in depth. For those who want to delve further into specific aspects then more extensive treatments can be found in books mentioned and described in the Reference section.

Maths underpins most of science, so we're going to begin by looking at some of its little known but truly fundamental problems.

1

Numbers Shmumbers: Why Being 'Bad with Numbers' May Actually be a Good Thing

The invention of the laws of numbers was made on the basis of error, dominant even for earliest times, that there are identical things (but in fact nothing is identical with anything else).

Nietzsche

Nietzsche was a German philosopher who died in 1900 perhaps from a brain tumour, or tertiary syphilis, or from an early form of dementia, depending on the source you choose to believe. At the age of 44 he suffered from a complete nervous breakdown that may have been precipitated by witnessing the savage whipping of a horse. After that he never really recovered his mental faculties.

So, he seems hardly worth quoting at all but many authors find themselves compelled to do so. This is because he was a dangerous and heretical thinker who was very good at producing rather terse and pointed quotes. 'God is dead' is probably his most famous; 'That which does not kill us makes us stronger' is another good one. Indeed, one whole book of his was given over to these aphorisms [1]. This plethora of quotable phrases makes Nietzsche irresistible to many authors because you can usually find one to support whatever case you are trying to make. So wide ranging are these scatter gun quotes that Nietzsche's work was at one point misappropriated by the Nazis, though he appears neither to have been anti-Semitic nor nationalistic.

But Nietzsche is worth quoting here because he was one of the first of the modern philosophers to harbour grave doubts about science and also about its most powerful tool: mathematics. Much of this doubt came from his own maverick nature but also from his early career as a classical scholar of Greek and Roman texts. Concerns about mathematics have been around for thousands of years but, eclipsed by maths' apparent success, they are often forgotten. These concerns had produced disbelief and even anger in some mathematicians but have, over time, been airbrushed out of its teaching so that few realise they constitute any problems at all.

In this chapter we will examine some fundamental problems with mathematics and how a lack of understanding of the innate unreality of mathematics has led to some of the most absurd concepts in science (Black Holes: I'm looking at you!).

For those who don't like or understand maths I want to reassure you that we're only going to use one simple arithmetical equation in this chapter. Lest you think I am being patronising to those who aren't 'good with numbers', I can assure you I mean quite the opposite, as you will see later.

Here comes the simple but extremely dangerous equation:

$$1 + 1 = ?$$

What's the answer?

Before we go any further, I want you to focus on the certainty you feel about knowing the correct answer. I'm going to take a wild guess and assume the answer you came to was '2'.

Well now I'm going to tell you the most awful truth about that equation. A truth that you never hear in schools or universities and that has, in many ways, been ignored for centuries. A truth that some might even argue is the root of all the evils in the world.

The truth is simply that, in the real world, 1+1 NEVER equals 2.

How can that be? Add one thing, say a soccer ball, to another identical soccer ball and you've got two soccer balls, surely?

But the fact is, as Nietzsche pointed out, that there are no two things that are identical in the whole universe. From grains of sand all the way up to galaxies: all have differences. Some of these differences are extremely subtle but the fact is they are still there. No matter how things are closely machined by nature (for example sand grains) or by man (for example footballs or sugar grains) there are always tiny differences.

Now you may consider this nit-picking. Grains of sand are so similar that it hardly matters that they are not identical. And for most of the works of man, for example shovelling sand into a cement mixer to help construct a building, that's true. But by making that simple approximation, by ignoring the leap we've taken for the purposes of simple expediency, we've disregarded what may be the most profound truth of all.

Science, in whatever shape or form, is the search for truth. All hard science, and much of the so-called 'softer' social sciences, is based on numbers. It's how we handle the universe. And all our equations and calculations are based on simple arithmetic. And that arithmetic is based on one plus one equalling two.

But, however you cut it, all our mathematics is built on a fundamental untruth because it supposes that two things can be considered identical.

Basing whole systems, whether they be scientific or political or whatever, on untruths can still work, for a while anyway. You could, for example, design a well lubricated mechanical system to run forever by pretending that friction didn't exist. It may run for a while but, sooner or later, any such system will fail. Nazi ideology was based on the idea that Germanic races were inherently superior to others. How did that work out in the end?

For quite a while now I've been putting forwards these arguments to physics colleagues and friends. This has elicited a range of comments and ripostes. Let me try and summarise these comments, leaving out the occasional rude word, and supply an appropriate response.

What about atoms? They're identical. One carbon atom plus one carbon atom equals two carbon atoms. Give me a break!

The truth is I don't know if atoms are identical and neither does anyone else because we can't see down to that scale. As far as we are aware they behave pretty much identically using our coarse methods of measurement. This might lead some to suppose they are identical but it certainly isn't proof. There are tiny differences in everything else we can see, so why shouldn't the same thing apply at the atomic scale? In earlier times, grains of sand, fleas and just about anything else that was tiny were thought to be identical until the microscope was invented. The finer the detail we can resolve, the greater the variation we inevitably find. The idea that atoms are really identical supposes a radical departure from our real-world experience.

It doesn't matter because maths works! Where would your iPAD and your smart phone and your Kindle be if it didn't? You're wasting my time.

Maybe. Maths does indeed work, within certain limits we will discuss later, and will continue to work up to a point. Then it won't, just like any other system based on a fundamental untruth. Maybe we should be at least cognisant that this approach sets fundamental limits on our perception of reality. Nothing is identical to anything else but in everything we do, from calculating population size to ordering a round of drinks, we ignore this. What are we missing by ignoring the one thing which appears to be universally true: that nothing is the same as anything else?

1+1=2 was never meant to be taken literally. It's about representation. One-dollar bill plus one-dollar bill may not give you two identical dollar bills because of slight variations in their physicality such as their weight (and not least because of the varying amounts of cocaine adhering to them), but what we are actually talking about is the representation of their worth.

Though a subtler response than the others, it is falling into the basic trap set by the equation in the first place. Financial value, corresponding to what a dollar represents is, like the identical objects required for arithmetic, a notional concept that resides only in our heads. It is not some hard, absolute, real world thing, though I can't help but accept that not having financial worth, notional or otherwise, can make life very real indeed.

The equation is only an approximation. Two grains of sand may vary in terms of their shape and mass but they are approximately the same.

That's a better answer but it only comes after you've pointed out that the equation is never actually true. Ask a hundred people what 1+1 equals and I doubt a single one will say: approximately 2.

And it is that word approximately that is so crucial here. Two objects can only be identical enough for the 1+1 to equal 2 if we determinedly ignore all the things that make them different. If, for example, we ignore the different shapes and masses and colours and elemental contaminants that go into each grain of sand. One Nelson Mandela plus one Joseph Stalin can indeed equal two humans but only if we ignore all their manifold differences.

The point is that even this most simple of arithmetic is an artificial construct that never directly corresponds to the real world. And yet this artificial way of thinking is drummed into us from an early age. It's one of the 'three Rs': reading, 'riting and 'rithmetic. Yet most teachers, themselves unaware they have taken this step into unreality (in other words treating some things as equal by the simple expedient of ignoring all the aspects of them that are different) do not pass on this awareness to the young people they teach in turn.

Perhaps this is why some people struggle with arithmetic and mathematics generally. These are often intelligent people who are perhaps labelled as 'artistic' because they are 'bad with numbers'. Maybe the reason for this is that arithmetic doesn't make sense to them at a fundamental level. Perhaps they are innately aware of the complexity of life where nothing is identical to anything else. Yet in their maths classes they have to pretend things are identical and will get punished if they don't toe the party line.

The problem for such people is exacerbated because they are taught arithmetic when they are only a few years old, when they are far too young to articulate any feelings of unease. Arithmetic is taught in a dogmatic and unquestioned fashion. How can a six-year-old stand up to that? Instead they retreat, tail between their legs, convinced they are somehow inferior because they are 'bad with numbers'.

As well as making perfectly intelligent people feel inferior, the nature of arithmetic, or perhaps the mind set from which it springs, arguably paves the way for some unfortunate consequences.

Early teaching of arithmetic is perhaps the first and most forceful way we are inculcated with the view that 'things' can indeed be regarded as identical. It is a powerful tool. After all, if in later life you own a pub then it is important to work out how many bottles of beer you need to order a month. It doesn't help you to focus on the slight differences in the shape and weight of the bottles and the minor variations in their contents.

The propensity to regard two objects as equal probably didn't originate with arithmetic, but instead may represent aspects of how our brains are hard-wired to work, as we will see later. A universe where everything is different from everything else is a scary place. If you were on the African plains ten thousand years ago, and were starving and needing to hunt an antelope, then you couldn't allow yourself to be distracted by the fact that each one was different. Instead, you needed to focus on the commonalities: for example, the footprints antelopes make, the way they move and so on.

For our ancestors, each animal didn't have a unique identity, they just become antelopes. It's how our limited, as opposed to omniscient, minds handle the notion of them. In a complicated universe one has to simplify to survive. There will, of course, be some nuances on top of this; each animal won't act identically and an experienced huntsman will factor this in when they track and hunt them.

As man developed away from hunting single animals and supplementing his diet by picking the odd misshapen fruit, he made the move towards agriculture. In doing so he began to deal with many animals such as chickens or cultivated vast orchards of fruit and vegetables. Selective breeding over time made all these animals and plants more and more alike and this made the need for developing mechanisms to handle these greater numbers even more compelling. Arithmetic was invented and the awareness of the non-identical nature of animals and plants began to fade away. What room is there for individual identity in a factory farmed chicken amongst a multitude of others?

And arithmetic spoke to something in the way our limited minds worked. An example is found in the matter of tribal identity. This concept allows us to deal intellectually with a ragtag bunch of wildly different individuals. If you want to warn a child not to go into the territory of a few hundred hostile individuals, you don't want to go through a list of all of them. It's easier to define them all as a specific tribe.

It's a handy, shorthand way of dealing with a complex situation but the problem is that by stripping out the complexity of individuality it starts to characterise all the people within each tribe as the same, especially tribes which aren't friendly.

And that is the first step on a very slippery slope. As soon as we start to mentally handle large groups of individuals by ignoring their differences, then we start to see them as 'all the same' and terrible things can happen. To Hitler, Jews were an undifferentiated mass that meant his earlier bad experiences with a few individuals tarred them all with the same brush.

Perhaps the earliest and clearest example of this linkage between arithmetic and evil can be found in the ancient concept of decimation. The word is often used wrongly nowadays to denote a massacre in which all, or at least many, of a group of people are killed. In fact, this was a technique originated in the early days of the Roman Empire. It was employed as a form of punishment for groups of people. The first recorded use was in 471 BC where decimation was applied to soldiers in a legion that had shown cowardice or had misbehaved in some way. Men were divided into groups of ten and drew lots. The person who drew the shortest straw was then bludgeoned to death by his colleagues. In other words, only one man in ten was killed.

No attempt was made to discern individual guilt or to discriminate in terms of any actions, good or bad, perpetrated by each individual. Instead they were all equated as 'the same' and their level of guilt was considered exactly equal.

Incidentally, though decimation suggests one in ten were killed, the number could be one in five or whatever was thought appropriate by the one doing the decimating.

Decimation did not begin and end with the Romans but was practised by the Italians during the First World War and the Soviets during the Second. Indeed, a Soviet general at Stalingrad personally shot every tenth man until his ammunition ran out [2].

And that's only what armies did to their own troops; decimating their enemies, such as captured prisoners, was also practised.

But the arithmetical 'quotas' which exactly equated one man with another and consigned them to death in vast multitudes, reached its pinnacle in Russia at the time of Stalin. It was known as the Great Purge

Russia faced huge potential unrest due to a famine largely caused by Stalin's forced collectivisation of farming. In order to tame this general disenchantment, Stalin developed an essentially random mechanism to keep Soviet citizens in a state of perpetual terror. Stalin's stated aim was to reduce the reservoir of terrorists and spies. Later on, it was used to reduce the threat from other wings of the communist party led by Bukharin and Trotsky.

Perhaps a million people were killed during the Great Purge or died due to the terrible conditions in the prison camps they were consigned to as part of other arithmetically determined quotas. Victims came from the top to the bottom of Soviet society, with five out of six of the original Politburo members succumbing, three out of five army marshals, eight out of nine admirals. Intellectuals of all persuasions were imprisoned and perhaps only a quarter survived. Peasants, churchgoers and clergy suffered similar fates.

At first, victims were targeted because of at least some suspected activities; for example, the study of sun spots was considered un-Marxist and nearly thirty astronomers paid for this with their lives. However, the situation became even grimmer when the Soviet leaders resorted to arithmetic alone. Top down calls provided the actual numbers to be executed within the military and across the regions of the country. Tens of thousands of executions were ordered without naming any specific individuals. Local party officials, to show their zeal and loyalty, sometimes asked for their quotas to be increased.

Arithmetic can clearly be a dangerous business, playing as it does to the limitations of how our minds work, so where did our interest in numbers come from and how did it lead to other aspects of what we call mathematics?

The History of Numbers

Counting has been around for at least 30,000 years, as the 55 marks in groups of five found on a wolf bone in the Czech Republic would seem to attest [3]. The grouping in fives is presumably because of the number of fingers on a human hand. Bearing in mind the material on which the marks were made, it's surprising the person had any fingers or even a hand left at all. That was one tough arithmetician!

Nowadays we're used to counting in units of ten and it was the Egyptians who started using this decimal system in about 3000 BC.

Pythagoras took numbers to a new level in the 6th Century BC. Though a Greek, he learnt his numbers from the Egyptians but took the whole business farther, transcending their basic use in counting. Numbers became sacred things in themselves. 'Number is the first principle', 'Number is the essence of all things', 'All is number' he wrote.

From being used to count chickens to becoming sacred is quite a jump for arithmetic, but this is only the first example of where something that was really only a tool has become the subject of veneration. As we will see, in the millennia that followed Pythagoras, scientists and mathematicians would often elevate the tools of their trade to the point of worship. The tools were the theories (mathematical, physical, biological) that tried to explain reality, but somehow in the process these often became that reality. The tools became the Laws.

Describing the history of numbers is dull stuff. Books that deal with it struggle to find much of humour so, when it comes to Pythagorus, authors inevitably focus on the issue of beans.

Poor old Pythagoras couldn't stand beans. Not only were they the cause of flatulence but he also thought they too closely resembled human genitals. No, I'm not sure why either.

So great was his aversion to beans that he would rather have had his throat cut than run across a bean field. And indeed, one day, chased by his enemies and finding his way blocked by just such a bean field, that is exactly what happened.

Pythagorus did other unusual things. He began his own sect and even, at one point, claimed he was a god. The sect was secret and new disciples weren't allowed to speak or to make any other noises during their first years of membership.

Pythagorus even had a man called Hippasus killed because he had the temerity to give away the most dangerous secret of Pythagoras' sect. I am going to explain what this terrible secret is so you'd better brace yourself for a major revelation.

In Pythagoras' perfect world of sacred numbers, any real number could be expressed as the ratio of two whole numbers. For example: three-and-a-half can be expressed as 7 divided by 2.

All was indeed perfect until Hippasus came up with his filthy heresy, namely that some numbers could not be expressed as the ratio of two whole numbers. One example of this is the pesky number pi (3.14159... and on and on).

If that's not worth killing someone for, then I don't know what is!

As we will sadly and repeatedly see, Hippasus was only the first in a long line of people to suffer and even die because they came up with an uncomfortable truth that did not conform to prevailing theory.

That's a terrible shame and waste because theory may never be the truth.

It would appear from Pythagoras' statements such as 'All is number' that he was perhaps the first one not to understand that numbers could only ever be an approximation to reality. Certainly Euclid, whose work followed on from Pythagoras, stated as his first 'common notion' that: 'Things that are equal to the same thing are also equal to one another.' This implies that, as far as Euclid was concerned, there were at least three identical things in the universe whereas the truth is that there aren't even two.

The Greeks thought all of nature could eventually be understood through mathematics and that all its workings could be unearthed by mathematical reasoning.

They're all long dead, and therefore can't sue, so I'm going to blame them for all the problems with science and mathematics described in this book.

That said, it must have been easy and comforting to be taken in by this way of thinking. If you'd never conceived of numbers then they might indeed seem like powerful magic. For example, any system which uses coinage is only possible by the rules of arithmetic. Numbers make many things possible and seem to put the world on a firmer footing. It's not surprising that some ancient peoples thought numbers held magical powers, often using them in their religious rituals.

The early Muslim world did a lot of thinking about mathematics and numbers (they weren't afraid of the irrational ones like Pythagoras) but then Muslim theologians pretty much stopped further development. The reason seems to be that they were concerned it would uncover secrets Allah might want to remain hidden.

It would seem that in the Roman, Greek and Muslim worlds there was a widespread belief in the absolute, and indeed sacred, meaning of numbers. Left far behind was the awareness that numbers were just simple tools that could only ever reflect reality in approximate ways.

Even in these supposedly enlightened times this is still essentially the case.

Before we leave the Greeks and Romans we should quickly discuss another aspect of mathematics that is neither an accurate reflection of reality nor something immutable and sacred, yet it is taught in the most dogmatic of ways. I am referring to Geometry.

Geometry

'There are 180 degrees in a triangle'. That's a mantra we all learn at school. It is also, in the real world, never true. Never ever. In the real world no triangle has perfectly straight sides. Whether it be tiny unevennesses in the paper we draw it on, or the pixelation on a computer's digital image, lines are never exactly straight.

In a sense this goes far deeper than inevitable tiny imperfections in our materials, at least according to what is now the conventional wisdom of General Relativity theory. This says that space is innately curved due to the proximity of any objects with mass, like planets or even just your iPhone. In space, and everywhere for that matter, there is no such thing as a straight line.

With no straight lines in the universe, there can be no perfect triangles with exactly 180 degrees. 'There are 180 degrees in a triangle' is therefore never true.

It's not that mathematicians don't know this but again, like Pythagoras making numbers sacred, they will instead talk of the perfection of their subject. For example, according to Berlinski: '...the Euclidean triangle, at once perfect and controlled, a fantastic extrapolation from experience, and entry into the absolute' [4].

The truth is that all we know of the universe is the imperfect reality we experience every day. All else is falsehood. Mathematicians have produced a fantasy. Sort of like Lord of the Rings, but with more utility.

So even our most basic mathematical concepts are flawed because they do not correspond to reality. That would be OK, just like Lord of the Rings is OK, if left confined to an imaginary world. Unfortunately, this imaginary world of mathematical perfection breeds concepts which bleed back into the real world. Things which are in fact phantasms are imbued with realistic con- notations and as a result spawn some of the most absurd notions in modern physics, as we will see in the next three chapters.

If anyone bases reality on any fantasy then, sooner or later, they are going to come a cropper.

Here are the first two of these mathematical phantasms:

Zero and Infinity ('maths' evil twins')

Zero is a concept that has wormed its way into our consciousness, though it was not always so. In fact, for most of mankind's mathematical history the notion of zero did not exist.

Man may have been counting on wolf bones, or whatever was to hand, for 30,000 years and yet never once had they had the need for something called 'zero'. After all, when you count your fingers you don't start with zero.

So where did zero come from?

Back in about 300 BC the Babylonians had something that could be con- fused with a zero, though in fact it was more like a spacer. Where we count in tens, Babylonians counted in sixties. So, for example, to differentiate 16 and 160 they might have 16*, where the * is the spacer.

So, the Babylonian spacer didn't actually represent 'nothing' as zero does nowadays.

Even the Romans didn't have zero which is why their calendar doesn't have 0 AD. Instead it starts at 1 AD, and that is why purists maintained that celebrations for the new millennium should have started as midnight ap- proached on December 31st 2000, not 1999 as everyone else thought. How- ever, the appeal to the modern mind of nice round numbers triumphed over this pedantry.

Zero also went against an axiom of Archimedes, the third century BC

Greek scientist and mathematician: add a number to itself and you get a different number. Of course, adding zero to zero just gets you zero so that would no longer work. Bearing in mind that 'axiom' means self-evident proposition requiring no proof then that would really be a spanner in the works if zero did exist.

It's also a bit unsettling that when you multiply any number by zero you always get zero, no matter how big the first number is.

It's like zero is screwing up the whole number system. No wonder the Greeks and Romans didn't like it.

It also raised fears of the 'the void' amongst the Greeks and, later, the Christians. God had created the world so the idea of void suggested somewhere where there was no God. This was frightening and indeed the void even became identified as the Devil himself.

Indian mathematicians, coming from a Hindu philosophy where everything sprang from the void, were however much more comfortable with the idea. Brahmagupta, the Indian mathematician, introduced rules for trying to deal with zero in 628AD [5].

Indians came to the concept of zero by way of negative numbers, themselves first mentioned in China in the Second Century BC. Negative numbers are far from being an obvious idea. In terms of flesh and blood reality, how can you have a negative ox, for example? You can perhaps have one less ox than you used to, but that's not the same as having a living, breathing negative ox. The only 'real world' meaning for negative numbers is essentially in regard to debt in one form or another, which again is only a human concept and has no absolute meaning in reality.

However, if you do accept the notion of negative numbers then suddenly you have a discontinuity, or break, between negative and positive numbers. The idea of zero filled that gap in the sense that it came between minus one and plus one.

Zero therefore came from the non-real-world concept of negative numbers. To illustrate just how non-real negative numbers themselves are, consider what happens when positive numbers are divided by negative numbers. According to the rules of arithmetic this leaves a negative number as a result. When pupils come across this for the first time they invariably dislike it because they can't equate them with everyday reality. I bet a few of you still don't like it today. After all, if you take six cars and divide them by minus two cars (what are minus two cars for a start?) you get minus three cars. Earlier on I suggested that, in the real world, negative numbers might be associated with debt but even this breaks down in this example. If you divide six cars you own by two cars you owe someone else, how do you wind up still owing three cars?

Perhaps in order to disguise what nonsense this all is, mathematicians have the temerity to define positive and negative numbers as 'real' numbers, perhaps in the same way that Orwell in his book 1984 called the bureaucracy for waging war 'The Ministry of Peace'. Orwell called this Doublespeak.

Getting a negative number when you divide a positive number by a negative number may 'fit' as far as the mathematics are concerned, but it doesn't 'fit' reality.

Nowadays mathematicians go even further in their fictions concerning negative numbers. The square root of minus one is central to some of the most important techniques they use. Indeed, I used one such technique myself nearly every day of my professional life as a physicist (it was the Fourier Transform but, don't worry, I'm never going to mention it again).

Even if you assume you can get a negative anything, then how can you find the square root of this? This is a number which, when multiplied by itself, can produce the negative result. Perhaps not surprisingly, mathematicians took to call these strange new things complex numbers. Rene Descartes made up another term, calling the actual square root of minus one an 'imaginary' number. Tellingly, this tag stuck.

Squaring any other negative or positive number, except for this imaginary one, gives a positive number. Even minus 2 multiplied by minus 2 is plus four. Yet multiplying an imaginary number by another imaginary number gives a negative number.

This weirdness led many mathematicians at the time to disbelief and even anger. Yet these determinedly non-realistic numbers found utility in certain mathematical processes which in themselves had major and often beneficial real-world impacts. For example, such mathematics is used in the processing of signals from everything from radio waves to the vibrations from seismic surveying.

This utility made it easy to paper over the rupture between the real world and the world of mathematics. Over time, negative and imaginary numbers became taught in schools as a stone-cold fact. No matter how greatly school kids were baffled by the concepts, no matter how divorced from reality they seemed, they had to learn them if they wanted to pass their exams.

But let's get back to zero. Zero means nothing. Certainly, in reality, there is no place anywhere in the universe where there is 'nothing', not even in the vacuum of space. No matter how far from the nearest star, space always contains something whether it is just a few atoms of matter or photons of radiation. In reality there is no such thing as 'nothing'.

Still need convincing about how strange zero is? If, according to the mathematics of zero, multiplying 1 by 0 gives you zero then, presumably, by then dividing this all by zero we should get back to 1 again.

Nope. According to our mathematicians it actually comes to zero.

It's almost as though, even within the framework of numbers (themselves unrealistic as they assume objects are identical) zero just doesn't belong.

Despite huge problems like this, Islamic scholars followed the Indian mathematicians' approach to accepting the concept of zero though it would take a lot longer for the West to follow suit. Indeed, it wasn't until the 13th Century AD that the West cast off the stranglehold that Greek scientific and mathematical thought had regarding the devil-like void (zero).

Western scientists began to believe that zero existed...

...or rather didn't exist, as zero is supposed to mean nothing.

The idea of zero is so strange that language, like mathematics, has difficulty coping with it and begins to break down.

Whether zero exists or not, it's a concept that percolates through all aspects of modern science, particularly physics, and helps engender theories that purportedly explain everything. A concept which itself may be entirely unrealistic, is widely used to explain reality.

It gets much worse, though. I hope you have found this section on zero mildly disturbing at the very least. If you have, then you'd better hold onto your hat when you read the next section because Pandora's Box really does swing open.

Infinity

There is one concept that corrupts and confuses the others. I am not speaking of Evil, whose limited sphere is Ethics, I am speaking of the infinite. Jorge Luis Borges in Otras Inquisitiones

The worst thing about zero is that, according to received wisdom, if you divide any number by zero you get an even more disturbing beast called infinity.

Nobody wanted the infinite. Right from the start Aristotle said that mathematicians 'do not need the infinite'. The reason for this is that in the real world nothing is infinite, a point which just about everyone today has forgotten because of habituation to the concept. There may be an awful lot of grains of sand on the beaches of the world, or atoms in the universe, but that isn't what infinity means. Infinity means without limit, or without end, or an indefinably great number.

Part of the hang-up about infinity, as with the concept of zero, came originally from religion. If God created the world/universe then before that there must have been that nasty devilish void. If you wanted to avoid that idea then you could instead believe that the universe had always existed which would mean it was infinitely old. From that point of view there is either infinity or there is zero, but it doesn't allow both.

Thus, concepts of zero and infinity are either inextricably intertwined or, from the latter view, mutually exclusive. The heretical view is that both are entirely fictitious.

Infinities and zeros, appearing like vermin in equations, send scientists through all sorts of contortions to make their equations 'work'. For example, when mathematicians divide zero by zero they have to believe they get the number one. These contortions spawn all sorts of outlandish concepts.

Infinity strains credulity even further because it needs to have its own rules otherwise it just doesn't work. For example, no other number can be equal to or greater than infinity (Galileo's rules). Divide infinity by any other number and you get: infinity.

Scratch the surface and the concept of infinity really doesn't make sense. For example, think of the infinity of even numbers. There will also be an infinite number of odd numbers. Does that mean there is a double infinity when you include both even and odd numbers. No, according to mathematicians there is just an infinity.

Many mathematical equations or expressions may routinely produce numbers that increase to infinity but that does not mean they have any correspondence with reality. Though habitual exposure to the concept has dulled our awareness of just how outlandish it is, in the same way as a lifetime of heavy manual work produces calluses which dull the initial pain, the thought of infinity is nevertheless still anathema to many mathematicians and theoretical physicists. Indeed, when infinity appears in physics equations purporting to represent reality, the first thought physicists have is that something has gone wrong. Generally, it leads them to consider that whatever the model representing reality is being used, it cannot be taking into account all the relevant factors.

Nothing in the universe that we can see is infinite. Even cosmologists don't believe the universe extends to infinity. Time might possibly be infinite but we don't know one way or the other. Even physicists accepting the present received wisdom of the Big Bang theory believe our universe 'popped' into existence out of a fluctuation in the zero-point energy (more on this later). This suggests that time is not infinite even under their belief system.

With no evidence that anything in reality is infinite, this has led to considerable dispute amongst scientists and mathematicians as to what infinity actually means. To try and bring this more under control, a mathematician called Cantor divided infinities into three types:

1) Mathematical infinities: these appear in equations and mathematical expressions. They are abstractions and only exist there and in the minds of men

2) Physical infinities: these actually appear in our universe

3) Absolute infinity: the sum total of everything

Famous mathematicians and philosophers have believed in different combinations of these, illustrating just how tricky or nonsensical this subject is (take your pick). For example:

Abraham Robinson	(Mathematical No, Physical NO, Absolute NO)
David Hilbert	(Mathematical YES, Physical NO, Absolute YES)
Bertrand Russell	(Mathematical YES, Physical YES, Absolute NO)
Kurt Godel	(Mathematical YES, Physical NO, Absolute NO)
LEJ Brouwer	(Mathematical NO, Physical YES, Absolute YES)
George Cantor	(Mathematical YES, Physical YES, Absolute YES)

(see references [6] and [7] for even more runners and riders in the infinity sweepstakes)

No wonder physicists are uneasy about infinities when they crop up in their equations. When they appeared in Einstein's work he shied away from them. These annoying infinities produced something called a 'singularity'. This is a point in space where gravitational forces would, according to the mathematics, cause matter to have infinite density and infinitesimal volume, thus making both space and time infinity distorted.

Einstein said of this mess: 'A singularity brings so much arbitrariness into the theory... it actually nullifies its laws. Every field theory...must therefore adhere to the fundamental principle that singularities of the field are to be excluded'.

Peter Bergmann, a collaborator of Einstein's, went on to say '...a theory that involves singularities...unavoidably carries within itself the seeds of its own destruction'.

Despite these warnings from Einstein, perhaps the greatest scientist of all time, other physicists and non-scientists take at face value these infinities and construct all sorts of esoteric phenomena out of them. These include Black Holes.

In engineering, where science meets gritty, complex reality, if an infinity appears then something is definitely wrong and it always means that some component, like friction, has been left out of the model or equation.

In quantum physics, where calculations often throw up infinities, physicists resort to splitting the calculations into finite and infinite components (see 'renormalisation' in the next chapter) then effectively subtracting out the inconvenient infinite components. As we will see, this is basically a fudge on a mind-bending scale.

So far, we've only been talking about the infinitely large, but what about the infinitely small? It must be said that the notion of something infinitesimally small has had great utility in a branch of mathematics called calculus. This involves integration and differentiation but, if you are faint of mathematical heart, don't worry because we are not going into those. This is a field which has found massive and successful application even though it is apparently based on nonsense.

All forms of calculus are based on 'infinitesimals'. These are numbers which are smaller than any other numbers but are not zero. These infinitesimal numbers can be added to themselves as many times as you like and yet they will still remain less than any given number. In other words, you can add as many infinitesimally small numbers as you want together and you still somehow end up with something that is infinitesimally small.

Already this isn't making sense and people have been arguing against it for centuries, even doubting Isaac Newton's calculus which is based on these beasts. Bishop Berkeley, an eighteenth-century mathematician and philosopher, argued that the calculus lacked rigorous theoretical foundations because sometimes their adherents used infinitesimals as positive non-zero numbers but other times as actually meaning zero. No wonder Berkeley called infinitesimals 'The ghosts of departed quantities'.

Newton, vying with Einstein for the title of Greatest Scientist of All Time, tried to gloss over this nonsensical aspect.

As Friedrich Engels [5] said when the infinitely small and infinitely large were introduced into mathematics: 'The virgin state of absolute validity and irrefutable proof of everything mathematical was gone forever, the realm of controversy was inaugurated, and we have reached the point where people differentiate and integrate not because they understand what they are doing but from pure faith, because up to now it has always come out right'.

And indeed, the attitude Engels was describing is still widespread in the scientific community.

Whilst many mathematicians objected to infinitesimals, more and more calculus was performed and with apparent success. Over time the fundamental nonsensical notion on which it was based became tacitly accepted. Another crack was slowly papered over.

But no matter how well something appears to work, if at its heart it is nonsense, then it should never be mistaken for the truth. Calculus has had a good run for three hundred years but it is based on assumptions which have no correspondence with reality.

The wallpaper may make the room look nice but it also hides nasty cracks in the supporting walls. Sooner or later the whole building may fall down and, judging by the insane concepts physicists have to resort to when it comes to dealing with the very large and the very small, it's already starting to crumble.

Not that it was ever on firm foundations to begin with as we will see in the next section.

Nature hates simple equations

When I was a student I found my physics textbooks greatly comforting. Like holy texts, they contained powerful knowledge revealing the workings of the universe boiled down by man's ingenuity into startlingly simple equations. Mechanics, thermodynamics, astrophysics, nuclear physics, the books had their principles pretty much nailed down.

Knowledge is power and that sort of power is enough to turn even a physics nerd into a world shattering colossus. The only trouble was that when I came to apply these equations professionally, I found that not a single one of them worked.

Ever.

Still in awe of the mighty intellects that had performed the experiments and developed the theory and written the textbooks, I naturally assumed the problem lay with me: that I had made a mistake.

I'm too embarrassed to tell you how long it took me to realise that it wasn't me being stupid, it was that the equations in the textbooks never accurately reflected reality.

Textbook equations generally take the form of one thing added to or multiplied by something else equalling a third thing. Unfortunately, the only way these effects and forces can be written down in such simple form is by ignoring a multitude of other factors thrown up by messy reality.

For example, if you tried to push an object such as a rock across the ground by applying a constant force then, from the mass of the rock and the constant force, you can calculate the acceleration you will give it. In Textbook

World, and only there, it works out as force divided by mass.

In the real world there is the obvious matter of friction between the rock and the surface over which it is moving. One would hope that including another term in the simple ('linear') equation would make it work to a point where it actually reflected reality. The trouble is that the effect of friction changes with the speed of the object, which is itself changing as it accelerates. The level of friction in turn affects the speed of the object in a kind of feedback loop. Suddenly the equation is no longer simple and has to include terms which change with time. The equation changes from linear to 'differential' and this is where calculus comes into it.

It gets even worse if we factor in air resistance.

This wouldn't matter if these non-linear equations could be solved. A few of them can be, but usually only under certain specific conditions. These no-brainer solutions are the ones proudly provided by the textbooks. What many scientists forget is that the vast majority of non-linear equations have no general solutions. Nature, in essence, is non-linear. In other words, it does not yield specific and unequivocal solutions to our mathematical analyses.

Computer simulations can be used to model what is happening but these are always based on such equations. Faced with messy non-linear equations with no simple solutions, the only way to get anything like a meaningful result is to make certain assumptions or approximations. For example, friction might be assumed to be constant over a certain range of velocities or, if the object was small, the effects of air resistance might be neglected.

Not only does this inherent inaccuracy limit the predictive capabilities of our models and equations but also, as we will see in the chapter on Chaos and Prediction, these equations are often highly sensitive to the starting conditions of the experiment. Tiny differences in these can have large and often unpredictable consequences.

Nature, in short, does not like to be constrained by simple equations. The intrusion of the innate complexity of the universe makes any attempt to predict it very limited and often completely wrong. Inherently flawed and limited mathematics can only ever take us so far.

So, maths juggles routinely with rather suspect ideas like zero and infinity and can never produce exact solutions when it tries to model the behaviour of the universe. Hopefully, though you may dispute some of the preceding arguments, you can appreciate the sheer artificiality of mathematics and even of simple numbers themselves. Yet throughout the ages, and despite the approximations and apparently nonsensical concepts which make up its foundations, many respected adherents of maths still believe that it generally reflects reality. Some believe it has even deeper meaning. Here are some typical quotes through the ages:

Philolaus, 4th Century BC: 'All things which can be known have number; for it is not possible that without number anything can be either conceived or known'.

Galileo, 16th Century AD: 'The Book of Nature is written in mathematics.'

Berlinski, 20th Century AD: 'The doctrine that number is the essential of all things, passing through the prism of a thousand philosophical tracts, remains the central insight of Western science, the indisputable key of coordination'

Seife: 20th Century AD '... calculus is the language of nature'.

Let's face this issue head on.

Is maths 'real' at all and, if it isn't, where does it come from?

That maths may have no intrinsic reality itself, that it may not actually underlie the workings of the universe, has been proposed before. LEJ Brouwer, a mathematician and philosopher in the early part of the twentieth century, was one such doubter and this prompted him to try to rewrite mathematics based on different, though sometimes rather vague, principles. He maintained that mathematics was only a social construction, being purely theoretical and not at all factual. Being simply a mental construct, it was therefore hobbled by the subjectivity and inherent limitations of the human mind.

Brouwer's fate is emblematic of what often happens to more contemporary heretics of science and mathematics. He became subject to fierce opposition with the mathematician David Hilbert becoming his bete noire. Overwhelming levels of criticism combined perhaps with something inherent in their heretical natures often cause these heretics to behave in odd ways. After years of bitter dispute, Brouwer developed a sense of persecution and became deeply paranoid.

Around the time of Brouwer, another philosopher and mathematician, the weighty historical figure of Edmund Husserl [8], was also disturbed by the way science and maths were progressing but for another reason. Until that point scientists had assumed they could be entirely objective when it came to observing experiments or phenomena. Even today when something is subject to 'scientific analysis' it is generally assumed that it will be treated impartially by unbiased observers.

However, Husserl was concerned that such objectivity was impossible. As human beings, everything we observe is filtered through a myriad of scientific, social and cultural assumptions, and is also affected by our limited abilities to sense things. For example, our eyes can see only a tiny part of the electromagnetic spectrum, our ears can hear only a narrow frequency range of sound, and this acts as a set of blinkers on what we can perceive and understand of the world.

If we take mathematics as an example, the mathematician comes to any problem with a series of implicit assumptions such as that there are always 180 degrees in a triangle, that lines can be straight and that 1+1=2 even though none of these are true in the real world. With these sorts of biases, how can anybody really be entirely objective about anything in the real world?

The objective observer of nature is, in other words, a fiction. This idea actually dates back to the philosopher Immanuel Kant in the eighteenth century who tried to distinguish what we could actually perceive with our limited sensibilities and mental processes (phenomena) and the objects themselves (he called these noumena).

This awareness of the inevitable subjectivity of the observer, though still largely ignored by the 'hard' science community, is utilised extensively in research into the social sciences as we will see in Chapter 12. When observing social groups, whether it is the homeless and destitute, or nurses working in a hospital, the observer will inevitably bring certain cultural and social biases.

Though Husserl was not against many of the ways of science or maths, he was concerned by what he called 'the mathematisation of nature'. This was the conventional way of thinking that somehow all the inevitable subjectivity, assumptions and value judgements of the observer could be set aside when investigating phenomena in the first place.

More recently, George Lakoff and Rafael Nunez [9], cognitive scientists at the University of California (at Berkeley and San Diego respectively) have tried for the first time to investigate the proposal that mathematics, rather than underpinning reality, was entirely a construct of the human mind. Their findings indicate that we are incapable of thinking in any other than certain hard-wired ways and that our view of the universe is therefore inevitably highly constrained. As we can't think outside the box (the box being our brains in this instance) then we have no way of knowing whether maths plays any part in how the universe actually works.

As cognitive scientists, Lakoff and Nunez investigated how the brain deals with mathematics and how it is able to perform mathematical operations such as arithmetic. Their works indicates that, contrary to the views of almost all mathematicians from Plato onwards, that maths does not transcend this blood and tissue limit. It therefore does not lie at the very heart of the universe.

And would this really be surprising? The mathematicians' belief in the transcendent power of mathematics is actually only an article of faith. Like the belief in God, we can neither prove nor disprove it. The thing we would need to use to construct any proof - the brain - is what we use to construct our theories in the first place.

So, rather than transcendent mathematics, we may only have 'mind-based mathematics'. Lakoff and Nunez came to the conclusion that mathematics 'is a product of the neural capacities of our brains, the nature of our bodies, our environment, and our long social and cultural history'. How did they arrive at this?

They investigated how many of our sophisticated mathematical concepts can arise from some basic cognitive hardwired abilities. This starts with an innate ability to tell the difference immediately between one, two or three objects which appears in infants within a few days of birth. This ability is called subitizing and may be the basis of our concept of numbers. Since these subitzed objects have an existence independent of the human mind, then this may lead to the notion that numbers have an independent existence too.

They also point out how much of our thought is based on metaphor. Rather than being a simple means of comparison, it actually makes abstract thought possible. For example, emotional attachment is thought of in terms of warmth, and words like warm, cold, icy are the linguistic currency for describing this complex thing.

For another example of how inherent metaphor is to the way we think, consider the concept of importance which is described in terms of size ('small matter', 'big issue'). We can only intellectually deal with abstract concepts like importance by resorting to these more concrete comparators.

One unfortunate consequence is that the use of metaphor itself introduces elements into the concept of importance that don't actually originate there. In other words, we become constrained to think of importance only in terms of the big and small even though it is, presumably, a far more complex and multi-faceted concept than that.

Lakoff and Nunez argue that mathematics is the laying of metaphor upon metaphor and this can result in a compounding of any inappropriateness, perhaps squeezing out important truths.

For example, the 'hardness' of the numbers 1, 2 and 3 that we get from our innate subitizing facility, gets into a nasty car crash when it meets the concept of infinity. If there are an infinite number of even numbers and an infinite number of odd numbers then how, according to mathematicians, can a combination of the two produce the same infinite number? Either infinity is nonsense or our numbers were never really 'hard' to begin with. Perhaps even both are non-realistic notions.

Hard numbers and infinity don't mix well, but what about geometry and infinity? That's not so good either, and both are in conflict with reality. Geometrically speaking, parallel lines never meet. However, mathematicians assume parallel lines meet at infinity, though it makes no 'common sense' (whatever that is). And in reality, one can never make or observe lines that are exactly straight and exactly parallel. They will meet, whether it is ten metres, or hundreds of thousands of light years away.

Perhaps none of this would matter if it was kept in the realm of abstract mathematics but unfortunately maths itself is used as a metaphor for reality. Infinities crop up in the maths underpinning physics theories that are then ascribed to an actual physical existence.

This use of metaphor, and the conflicting meanings it produces, can also be seen as regards the concept of zero. If we are counting, say, clothes-pegs in a bag the metaphor we are using is the idea of collecting objects. In this instance then zero comes to mean emptiness. If we start moving, then zero becomes the point from which we start (the origin). If we are measuring something, for example with a tape measure, then zero means ultimate smallness. If we are grouping objects together then zero means a lack of. Thus, depending on the metaphor we are using, zero can mean emptiness, nothingness, lack, absence, ultimate smallness or origin. How we think of it affects what we mean by zero. Can the so-called perfect mathematical world therefore ever be divorced from our flesh and blood thought processes?

Incidentally, no matter what metaphor you use, dividing by zero produces no sensible result. No wonder we all have trouble conceptualising such a process.

Another example of how subjectivity has inevitably crept into mathematics, though perhaps more from cultural bias than flesh and blood hardwiring this time, can be seen in the apparently unquestioned need to use the smallest number of axioms (a self-evident or universally recognised truth) in any branch of mathematics. For example, in geometry Euclid said that only five such axioms, which he considered actual 'truths', could explain the whole field of plane geometry. This belief has held fast for over two thousand years.

However, even if maths has some reality there is no clear reason why this minimisation is necessary. Yet this need for minimisation is unquestioned in Greek mathematics and it is upon this that the now predominant Western mathematics is built. Such a concept wasn't felt necessary in Egyptian or Babylonian maths.

That all of science and maths will ultimately be broken down into a minimum number of truths is an almost universally held belief by scientists and mathematicians, yet why should it be so?

This assumption that science and maths should be based on a minimal set of postulates has been described by Lakoff and Nunez as the 'folk theory of essences' and is one of the inherent notions in what they call the 'Romance of Mathematics'.

The essential points of this Romance of Mathematics, to which the great majority of mathematicians subscribe, are that:

Mathematics is real, universal, absolute and true.

Mathematics would be there in the universe with or without man; in other words, it is transcendent.

The language of mathematics is the language of nature; that mathematics lies at the heart of physical phenomena. For example, that planets move in elliptical orbits because the relationship between the planet and the star it orbits is underpinned by laws which are governed entirely by mathematics.

Compelling though this all sounds there is, as our Heretic would point out, no proof to support any of these notions. They are, quite simply, articles of faith.

Instead, mathematics may be actually only the product of cognitive processes within the human brain whose perceptions of the world are limited by its senses. The brain's thinking, hardwired by our neural circuitry, is based on metaphor and analogy, our understanding of which is in turn limited by our flesh and blood experiences.

'Laws' of physics are stated in mathematical terms and the universe is supposed to 'obey' these, yet our Heretic would say that all we are seeing are certain regularities in the behaviour of matter. Mathematicians and physicists fit our brain grown mathematics to describe these regularities. Sometimes the fit may be good but that does not make it a 'Law' that the universe must obey.

Our Heretic has another question. If maths really does underpin the universe and is 'real' and is not just some human brain grown concept, then why do so many of its concepts have no real-world existence or even meaning? In the real world where is a negative number, never mind the square root of one of these? Where is there anything that is infinite? And, of course, negative numbers and infinity hardly scratch the surface of the many strange concepts in maths that are not even considered by mathematicians to have an analogue in the real world. For example: transcendental numbers, surreal numbers, hyper-real numbers, transfinite numbers, quaternions, empty sets and so on. If even mathematicians accept these aspects of maths have no direct correspondence with reality, then why should any of it?

Both our Believer and Sceptic find their voice at last and their rebuttal is a good one. It was an argument raised in the title of a paper by Nobel Prize winning physicist Eugene Wigner [10]: 'The unreasonable effectiveness of mathematics in the natural sciences'. Wigner concluded: 'The miracle of the appropriateness of the language of mathematics for the formulation of the laws of physics is a wonderful gift we neither understand or deserve'.

They both point an accusing finger at the Heretic. 'If maths isn't real then how come it works? Our whole social system would collapse without numbers and we'd be grubbing around in the dirt for roots to eat!', they roar lustily.

The Heretic replies that perhaps the reason our maths very often works in, more or less, solving problems is really because we have a whole range of mathematical techniques we can try fitting to the observations we make. Even if none of these produce a reasonable match, then we freely invent new mathematical concepts and models and methods. We fit our models to the experimental facts but that does not mean the universe works because of this maths. A ball falling under gravity does not make loads of calculations to take into account its mass, the mass of the Earth, its distance above the Earth, friction with the air and so. It falls because that is what it does. Humanity tries to fit models, created by our brains, to explain all this. Even then we will never get the results exactly correct because of the universe's complexity.

So, in summary I hope that, whether you accept our Heretic's view or not, that you at least come away from reading this chapter with an appreciation of what may be the sheer artificiality of mathematics.

Many mathematicians seem to view maths as a perfect world in itself. Our own existence, with its mess and fuss and noisiness, appears to them as a rather poor offshoot. Reality is the Hell to Maths' Heaven. Maths to them is how things should work because of its transcendence and beauty.

The Believer, of course, buys this unequivocally. The Sceptic holds his hand out, palm flat, and waggles it from side to side a bit. Maybe.

The Heretic, meanwhile, reaches for his screwdriver. Maths to him is a tool just like that. The screwdriver has a function which it can do well under some circumstances. However, we don't elevate this humble tool to the point where we believe in a sublime Screwdriver World where the Screwdriver, in its perfection, can unlock the secrets of the universe. A World compared to which our humble reality is simply a pale and tawdry reflection.

Maths, like the screwdriver, is just a useful tool that only works in context and within certain limitations.

And if you're 'bad at maths' then perhaps, rather than representing some mental limit on your part, it is because you cannot reconcile its simplistic notions with the complex reality that you daily experience. Why should you suddenly make the leap of faith required to elevate what is a simple, and at best approximate tool, into something worthy of total and unquestioning belief?

Even basic numbers may be suspect in that they assume that there are things in the universe that are identical. In fact, the one absolute truth we can see with our own eyes is that nothing in the universe is identical with anything else.

It may be that the maths we use is simply a product of our flesh and blood cognitive processes and may, by limiting us in this way, prevent us from ever grasping how the universe actually works. Even if we weren't blinkered like this, according to the Heretic we would still never understand how the universe works. This is because he believes it does not follow laws or principles or rules at all. The universe just carries on serenely no matter how many convoluted equations we devise to 'explain' it.

And all this is troubling as so much of the hard sciences like physics are based on these shaky and unrealistic mathematical foundations.

But a final reassurance for those of you who are 'bad at maths': though maintaining our distrust of the concept of infinity, we will not be dealing with maths again in this book.

Now we are going to look at the 'hard' science of physics. We'll start with physics on the big scale. Cosmology is the study of the universe and we will see how many desperate convolutions that science has had to go through to try to explain what the universe is made of, how it came to be, and what will supposedly be its fate.

2

Science of the Very Large

Scientific models are representations of reality, not reality itself, and no matter how well they work or how accurate their predictions under appropriate circumstances, they should always be regarded as approximations, and aids to the imagination, rather than the ultimate truth.

John Gribbin [1].

If only all physicists were as careful in what they say as the astrophysicist and science writer John Gribbin.

All physics, indeed all science, is essentially about models. Models are 'thought constructs' to explain why the universe works the way it does. Such models are usually underpinned by mathematics which, as the last chapter has tried to show, is itself a thought construct with perhaps only a tenuous relationship with reality.

That the models used can never reflect the absolute truth is something that many scientists do not realise. To them, the laws of physics and of the other sciences are stars to steer by. In the words of Scotty from the Star Ship Enterprise (an engineer rather than a scientist, admittedly): 'Ye canny break the laws of physics'. Yet the laws of physics are broken all the time whenever unexpected new data appear. They are then unceremoniously discarded to be replaced by other 'unbreakable' laws.

In this chapter we will discuss cosmology, the study of the universe: what it is, where it came from and where it's going. We will see the twists and turns that cosmologists have had to go through over the centuries to explain a very strange universe.

Trying to understand the cosmos is a difficult business because everything in it seems to be so far from everything else. Without instruments like telescopes to provide more information, ancient cosmologies were generally based on religious or spiritual beliefs [2]. The earliest cosmology we know much about is Babylonian, dating from about 3000 BC. This held that the

Earth was a flat disk and the sky was a solid bowl above it. Outside the bowl was something interpreted to be water or chaos or both.

Hindu cosmology originated about 2000 BC and was based on huge cycles of time. This estimated the cycles the universe went through as being about a few billion years in duration. This was a colossal number to even imagine four thousand years ago. Interestingly, modern cosmology would have us believe the age of the universe is about thirteen billion years. We will see that as cosmological margins of error go, that's a gnat's hair away from a bulls-eye.

Besides our own universe, the ancient Hindu scientists also believed there were innumerable others, at least according to a later text, the Bhagavata Purana, dating from perhaps 1000 BC. That might sound crazy, but today's physicists also resort to multiple universes at the drop of a hat in order to explain some of their experimental results.

Finally, the Hindus believed there was yet another grand cycle involving all of existence and they put the duration of this at over 300 trillion (300 million million) years. This greater cycle would be repeated forever.

Biblical cosmology appeared around 500 BC and was similar to the Babylonian with the dome of heaven keeping out the chaos ocean. Rain and snow were kept in storehouses outside this firmament and were let in through windows in the dome.

Leaving aside the cynical view that science is itself a religion full of articles of faith for which there is no evidence, arguably the first non-religious based cosmology came in with some Greek philosophers in around 500 BC. These were the Atomists and they believed the universe was filled with infinite nothingness except for an infinite number of small particles called atoms. Everything happened by basic cause and effect and nothing was random. This was the first really mechanistic view of the universe and finds clear echoes in the science of today.

Later in Greece came the development of the Pythagorean Universe around 400 BC. This had a fire in the centre of the universe and around this both the Sun and Earth orbited.

Whether the Sun, the Earth or something else was the centre of the universe would be a bone of contention in opposing cosmologies for the next two thousand years.

From the Stoic version of cosmology dating from around 300 BC came the idea that the universe was an island surrounded by infinite void. This universe went through cycles in which its size changed.

Around 280 BC Aristarchus, a Greek astronomer, held the view that the Earth and stars revolve around the Sun. This was in contention with both the Ptolemaic and Aristotelian views which put the Earth at the centre of the universe, with everything else revolving around it.

Much later, the Hindu idea of multiple universes made a re-appearance in ideas proposed by a Sunni Muslim theologian, Fakr al-Din al-Razi, in 1200 AD. He rejected the idea that the Earth was the centre of the universe and proposed an infinite number of universes. Despite this idea dogging science to this day, there is still absolutely no evidence that this is the case.

However, by now in the West the Christian view had become ascendant and Earth was very much considered to be the centre of the universe. Religious based, it meant that views to the contrary were viewed as heresy. And when it came to heretics, the Church at that time could be extremely unforgiving.

This Christian cosmological view was to hold for well over a thousand years until Copernicus came along in 1500 AD. He proposed that the Sun was the centre of the universe.

That was a very brave thing to do.

Nicolaus Copernicus was a remarkable man. A Prussian, he had shown considerable talents in mathematics and the formulation of theories of economics. He was the first to propose the quantity theory of money which said that the total amount of money available in an economy was directly related to the price of goods.

As well as all this, Copernicus also found time to operate as an unaccredited physician and a diplomat.

Copernicus had been instructed by the Catholic Church to reform the Julian calendar to ensure the timing of its rituals were correct and consistent. Copernicus attempted to do this by using the motions of the stars and the planets as a reference. Unfortunately, his observations and deliberations led him to the controversial view that the Earth orbited the Sun. This set Copernicus on a collision course with the Bible, which contained assertions such as in Psalm 104:5: 'the Lord set the earth on its foundations; it can never be moved'.

That Copernicus would eventually die a natural death despite this apparent heresy is a testament to his diplomatic skills. Though the theories he was proposing were contrary to church doctrine, he was careful to hedge his bets by dedicating his book on the subject (De revolutionibus orbium coelestium) to Pope Paul III.

Copernicus got away with it perhaps because what he was saying was still dismissed as just another theory. However, a hundred years later Galileo Galilei was brave enough to really put some observational meat on the bones of Copernicus' heretical idea.

Galileo came to the conclusions that planets orbited the Sun based on his observations of the phases of Venus using the telescope he had developed from spyglasses originally invented in 1608 in the Netherlands. Instead of spying on enemy positions, Galileo was spying on the cosmos.

Amongst other phenomena, he observed moons orbiting Jupiter which brought the realisation that moons could still orbit around other celestial bodies which were themselves moving. In turn this made him think of the moon orbiting the Earth. The Earth could therefore still be orbiting the Sun, just as Jupiter did. This is called the heliocentric view where the Sun is considered the centre of the universe.

The rejection of both the Christian and the much older Ptolemaic/Aristotelian views of the Earth being the centre of the universe was something that did not sit well with the Church. Indeed, Galileo could have lost his life had he not recanted his heliocentric views when he came under investigation by the Inquisition. Even then he was put under house arrest for the rest of his life.

Galileo's persecution is taken as damning evidence of the heavy hand of religious orthodoxy suppressing science, yet there are contrary views [3, 4]. Rather than being dogmatic itself, the Church may in fact have been objecting to Galileo's dogmatism that the Sun was the centre of the universe. The Church may have simply been trying to get Galileo to state that his heliocentric approach was only one model of planetary motion, rather than some absolute truth.

If so, then this is ironic because modern science presently takes exactly the same view as the red-hot iron wielding Catholic Church in the 16th century, namely that the Sun is not the centre of the universe.

There is certainly a lot of misapprehension about this whole incident. Galileo himself was hardly the heretical firebrand that he is sometimes portrayed and was, in many ways, rather pious though he did sire three children out of wedlock. He was also not the clear and unbiased thinker we sometimes believe. His heliocentric views led him to propose an erroneous theory to explain the tides we see on Earth. That his theory only produced one high tide a day, whereas there are two almost everywhere on Earth, was something that Galileo dismissed as an anomaly. He may have been the first, but was certainly not the last, scientist to dismiss inconvenient observational data in this way.

Galileo had no truck with the astronomer and mathematician Johannes Kepler's proposals that the moon was the primary cause of the tides though this is nowadays considered to be the truth. Also, despite Galileo's own careful observations of the planets, he never accepted Kepler's views that planetary orbits were ellipses. Instead he went along with the orthodox view of the Church that the orbits were perfectly circular.

He also rather brutally rejected the proposal of Grassi, a Jesuit monk and Professor of Mathematics, whose observations of cometary motion led him to believe that comets were further away than the moon, though Galileo presented no proof to support his views.

Despite much of what he proposed being incorrect, Galileo is considered one of the first modern scientific thinkers. Indeed, you can find much of modern scientific orthodoxy in some of the things he said, for example: ' [the Universe] is written in the language of mathematics, and its characters are triangles, circles and other geometric figures'.

As Einstein said of Galileo: 'Propositions arrived at by purely logical means are completely empty as regards reality. Because Galileo realised this, and particularly because he drummed it into the scientific world, he is the father of modern physics --indeed, of modern science altogether.'

Thus, Galileo and his Renaissance contemporaries like Kepler are general-ly considered as starting the move away from the study of the heavens, celes-tial mechanics, towards the study of a universe they believed was physically based. Kepler's observations of planetary motion, rather than using religious text as a basis, were based entirely on observation.

Condensed histories of science usually give the impression of clarity of thought in a few individuals like Kepler that allowed them to rise above the religious or social dogma of their day and thus push forward the march of objective science. As ever, the truth is much less clear cut. Like Galileo, Kepler was also a religious man. Despite his discovery that planets followed elliptical rather than circular orbits, he was still in thrall to the idea of spherical forms. For example, in Mysterium Cosmographicum he said 'the image of the trine God is a spherical surface, i.e. the Father is the Centre, the Son is the exter-nal surface and the Holy Spirit is like the rays that from the centre irradiate towards the spherical surface'. He also believed that planets, like men, had souls. Also, if your mind is still smarting from the idea mentioned in the last chapter that parallel lines meet at infinity and you're wondering who to blame: Kepler is your man.

Mankind's desire to impose order on the universe meant that a theory was required to explain the more or less elliptical orbits that Kepler deduced from the observations of the astronomer Tycho Brahe. And when it came to laying down the laws of physics, Isaac Newton was the greatest authority of the day. Newton, a devout 17th century Christian and alchemist, used his ideas of universal gravitation to construct laws that explained Kepler's findings. Indeed, these laws very successfully calculated planetary orbits, allowing predictions of where planets would be found at any given time set against what was then considered the immovable background of the stars. It was this predictive success that first led to the idea of fundamental laws governing the universe. It became the goal of physics to unearth these. After all, if man could work out the motions of the mighty celestial bodies then, surely, he would one day understand everything by characterising physical reality in mathematical terms.

However, at their heart, Newton's laws of gravitation did have a rather scary and fundamental weakness. Gravitational force between two objects is, according to Newton's Law of Universal Gravitation, the product of their masses divided by the square of the distance between them. In other words, if you double the distance between a planet and a moon, say, then the gravitational attraction is reduced to a quarter of the previous value.

That's all well and good but how does one mass 'know' how fast to fall towards the other? Can it be that by some spooky 'action at a distance', one object is somehow aware of how massive the other object is and also how far away it is? Does one body somehow send out this information in a kind of signal to the other? And if so, how does this signal propagate through space? How quickly does it travel? If one object was at the opposite side of the universe to the other, how quickly would it be aware of its gravitational attraction towards the other?

These were questions which troubled few. After all, the law worked in terms of its predictive capability. Whilst Newton's law gave a good estimate of the force of gravitational attraction, it didn't actually explain how it worked.

Mankind's quiet satisfaction gained by apparently divining God's Laws in mathematical form lasted for several hundred years and, apart from a slight discrepancy in the observed orbit of Mercury which didn't match the Law's predictions, everything appeared to be in order.

As we now move into the twentieth century the whole business of cosmology seemed pretty much worked out but this was just the lull before the storm. However, the turmoil of the previous two thousand years, where religious based theories of cosmology had vied with one another, was nothing compared to what was to happen in the twentieth century.

As late as 1900 it was assumed the universe was constant in size and consisted of just our galaxy. Only in 1920 did the astronomer Edwin Hubble first suggest that some 'clouds of dust' in the heavens were actually other galaxies. Though this was one more blow to the idea of man not being at the centre of the universe, by now the grip of the Church had greatly loosened and Hubble did not come even close to being tried as a heretic.

Perhaps emboldened by not being burned at the stake, Hubble then went on to make another seminal discovery in 1929. He found that the galaxies were apparently receding from each other. His observation depended on something called the Doppler Effect. As objects move, the light they emit is frequency shifted (in other words its colour changes) depending on the speed. The faster a light emitting object is moving away from us, the more the light is shifted to the red end of the spectrum. If it were moving towards us its light would be shifted to the blue.

This, incidentally, is how cosmologists come to the conclusion that galaxies are rotating about their axes. Lens shaped galaxies which happen to be positioned end on to us will have one side where the light is red shifted whilst the other is more shifted towards the blue.

Meanwhile, Albert Einstein, troubled by Newton's 'action at a distance' had produced his General Theory of Relativity. As well as providing new equations of gravitation that could accommodate the discrepancy in Mercury's orbit, his theory also brought with it a conceptual framework that explained the mechanism of action at a distance. Einstein proposed that mass warped space. The simplest analogy is to imagine a rubber sheet stretched taut. Put a heavy metal ball in the middle of the sheet and it will make the previously flat surface curve down to form a well (a so-called gravity well). Put another, lighter ball on the edge of the sheet and it will roll down the curved surface until it comes into contact with the central ball. Objects are attracted to each other because of the curvature of space produced by their masses.

And there is some evidence for this. As space itself is bent by mass, then the path of light through it is also bent. For example, when the sun crosses between us and a far-away star, the light from the star is deviated in a measurable way.

Though that sounds convincing, it should be pointed out that Newtonian physics also predicted the bending of light, though to a lesser degree [5], but nevertheless this discrepancy was enough to finally end the two-hundred-year reign of Newtonian gravitation.

Einstein had produced a compelling explanation that told a story by using a nice analogy. As previously described in the chapter on mathematics, the way we think may be greatly based on analogy. Thinking by analogy may be powerful but it does set limits on how we think and as a result may not always, or perhaps even ever, represent the actual truth. There's clearly a difference between an essentially flat, two-dimensional rubber sheet being stretched into the third dimension that is depth, and a three-dimensional space somehow being warped into another dimension we can't even perceive. But perhaps it goes further than that; something else entirely may be happening here which, mired in analogy that we are, we just can't begin to grasp.

It wasn't just Einstein's General Theory of Relativity that would impact on our understanding of the Cosmos in profound and unsettling ways. Einstein also produced a Special Theory of Relativity concerning the effects on objects moving at high speeds relative to one other. This too has also found some experimental verification.

Einstein's theory came about because he made one central assumption, namely that the speed of light was the same wherever you were and however fast you were travelling. This was, and still is, counterintuitive. If you are moving towards something that is emitting light then presumably the light reaching you will have a velocity which is the sum of its own velocity plus the velocity you are moving with.

This seems not to be true at all. In 1887 Albert Michelson and Edward Morley had set up an experiment where they sent out beams of light at 90 degrees to each other. They believed space was permeated by something called aether which light moved through. Again, we may have a situation where we're getting hung up on analogy. Sound can't travel unless it has air to go through. Water waves can't travel unless they have...well, water to travel through. This supposedly indicated there had to be something that light travelled through and it was this that was called the aether.

Michelson and Morley reckoned that as the Earth was moving round the Sun, then the light beam that was moving closer to the direction the Earth was moving would wind up moving faster than the beam at ninety degrees to it. However, no matter what orientations they used, and however often they tried it, the beams were measured to be travelling at the same speed. This seemed like a crazy finding. If science's faith in the underlying mathematical clockwork of the universe were to be maintained, then some mighty strange concepts would be required to explain this away.

Einstein's bold surmise was essentially to ditch the analogy of light with sound, saying instead that light travelled at the same speed whichever direction the object was moving. It just did. However, when he threaded this into the equations of moving objects it produced startlingly counter-intuitive results.

According to Einstein, when an object reaches high velocities with respect to an observer, strange things happen. To the observer, the object will appear to contract in the direction it is moving. In other words, if it started out as a sphere like a football, the faster it moved the more rugby ball shaped it would become. Also, if the observer could somehow measure it, he or she would find the mass of the object was also magically increased.

Not only that, but if there was a clock on the object that the observer could see, then it would appear to be running more slowly. In other words, the object's time frame would be slowing down compared to the observer's.

It's worth mentioning that all these surreal effects on the object being noted by the observer would not be experienced by anyone on the object itself. Everything would appear quite normal to them.

Now all of this would have been dismissed as madness (indeed, if this is all news to you then maybe you are feeling the same way now) had some of these effects not been experimentally observed.

For example, that mass increases with velocity has been observed in particle accelerators around the world. Indeed, as these objects approach the speed of light then greater and greater energy is needed to make their enormously increased masses move faster.

And it is at this point that our first nasty infinity emerges from the equations. As the velocity of an object, whether it is a sub-atomic particle or a spaceship, approaches the speed of light, Einstein's equations indicate that the mass of the object becomes infinite. Getting it to accelerate faster would require, again according to the equations, an infinite amount of energy. It is because of this that the Special Theory of Relativity precludes faster-than-light travel.

Einstein was arguably the greatest scientific mind there has ever been. So brilliant was he that, stunning though his work on Special and General Relativity was, he actually received his Nobel Prize for something else entirely, namely explaining a phenomenon called the photoelectric effect.

It's at this point that our Sceptic and Heretic really start to squabble. The Sceptic would point to this apparent experimental corroboration of increasing mass with velocity, as being clear evidence that Einstein's theories really did reflect and, to a degree, 'explain' reality.

The Heretic meanwhile, would point out that Einstein's was only one of many theories formulated to explain Michelson and Morley's experimental findings. The predictions of these other theories weren't corroborated by subsequent experiment and so fell by the wayside. But the Heretic points out that that doesn't mean that Einstein was correct, but rather that his theories survived because they were apparently the fittest, just as Newton's theory of gravitation had been the fittest for several hundred years before this.

Einstein was only the first to resort to really strange ideas to explain data from experiment, but his work freed physicists from feeling they needed to comply with 'common sense'. This, as we will see, opened a Pandora's Box and in time new theories would be proposed that would make Einstein's bizarre and revolutionary ideas appear tame.

Indeed, the Heretic would take the view that this resorting to outlandish ideas, with no grounding in our human experience of reality, is actually a sign of desperation; that scientists are trying to explain a universe which is actually inexplicable.

But leaving aside our grumpy old Heretic, there are other worries about Einstein's work. Its central tenet is that the speed of light is constant throughout the universe. In order to make this the case then moving objects have to have their mass increased, their time frame slowed and their length shrunk. Why the speed of light should be the fulcrum around which the whole mechanism of the universe essentially revolves is not really explained, just as Newton couldn't 'explain' the spooky action at a distance on which his theory depended.

For something so fundamental and otherwise unchangeable, it's also rather strange that the velocity of light varies depending on what it is travelling though. In vacuum it's nearly 300 million metres a second, in glass it travels about 88,000 metres per second slower. The Heretic wonders why a pane of glass can change the speed of light but the universe can't.

The apparent constancy of the speed of light, at least in a vacuum, has therefore become one of those physical constants like the gravitational constant (still called the Newtonian gravitational constant) beyond which science will go no further. The Heretic's view that the universe 'just is' may seem a cop-out but when it comes to the so-called universal constants then physicists do exactly the same thing. The fundamental universal constants are where scientists also have to hold up their hands and say that they 'just are' and no further explanation for their centrality to existence is possible.

One line of more contemporary reasoning, the Anthropic Principle, has it that without the universal constants being exactly what they are then life would not be possible. If the constants were different then atoms would fly apart or collapse into each other; stars and planets could never form. There would therefore be nobody around to ask why the constants took those particular values.

Some see this somewhat circular argument as being an attempt to put man, or at least human consciousness, back at the centre of the universe just as he was in pre-Copernican days. We will see more of this human centric view when it comes to the role of the observer and how it supposedly affects quantum events in the next chapter.

The Anthropic Principle, in any event, does not explain why any constants are required to make a universe in the first place. It just says they have to have certain values for us to be able to exist to observe them.

But now our historical review is approaching modern times and we are coming to the theory that is the current Top Dog at trying to explain the universe.

The Big Bang Theory

Hubble's observations seemed to show that the universe was expanding and cosmologists had to make major changes to their theories to explain this. If it was expanding then they inferred that it had to have expanded from something, perhaps one big ball of mass and energy called the ylem. All the particles in the universe would have emerged from this and would have gone on to form elements and molecules and suns and planets and us. Physicists figured the ylem came into existence immediately after something they called the Big Bang.

Taking the measured rate of expansion of the universe and then extrapolating back, they estimated the age of the universe as being about 7 billion years. However, in the early 1970s, new estimates were made and, suddenly, the age of the universe was put at 13.75 billion years. This overnight doubling of the age of the universe produced little public interest or alarm, showing just how much things had changed since medieval times.

The Big Bang Theory has it that all the matter and energy of the universe just suddenly popped into existence. Nobody in the early days of the theory liked this idea. A prejudice against something so big coming out of nothing is deeply ingrained and can be traced back to the early cultural ideas that underpin a lot of modern Western science. Aristotle and most Western scientists up to the start of the twentieth century did not believe in the idea of a void at all, but that is just what there must have been before the Big Bang.

Though the Big Bang itself seems pretty much accepted nowadays, it wasn't always the case. In fact, the term 'Big Bang' phrase was coined, facetiously, by the astronomer Fred Hoyle who didn't believe in it for a second. Hoyle had another theory, based originally on the ideas of Sir James Jeans in the 1920s, called The Steady State Theory. For a while this had a similar popularity to the Big Bang Theory and held that new matter was created to fill gaps left by an expanding universe. In this theory the universe had no beginning and no end.

The Steady State Theory fell back into obscurity because of the microwave radiation which perfuses through the universe. This background radiation appears to be very uniform. Scientists assume that the cosmic microwave background is left over 'relic radiation' from the Big Bang. High levels of uniformity of this were predicted from the Big Bang Theory.

This was enough evidence to sink the Steady State Theory but it is nevertheless based on a number of assumptions. For a start, scientists assume they understand how particulate matter evolved from this Big Bang, even though it all happened billions of years go. They also assume that space itself formed in the Big Bang and expanded from that point, thus affecting the wavelength of the microwaves. In other words, there wasn't even a big empty space before the Big Bang that the Big Bang exploded into and started to fill.

So basically everything, including space itself, came from absolutely nothing. That's one mighty big assumption.

And, of course, as the Heretic gleefully points out, it's all nonsense. Nowadays cosmologists would have us believe 95% of the universe is Dark Energy and Dark Matter, not one iota of which was predicted from the Big Bang Theory.

Of course, multitudes of cosmologists are working to reconcile (or 'fudge' as the Heretic would cynically have it) the present theory with these new ideas. If you think that would be a huge 'reconciliation', then I am afraid there is far worse to come in this chapter and the next.

That the Big Bang Theory predicted the microwave background pales to insignificance before this other massive failure of prediction. As we will see time and again in this book, elaborate theories can make apparently correct predictions for centuries but it does not mean they really reflect or explain reality.

And that's not the only reconciliation that the Big Bang Theory has needed. The idea that the universe grew steadily from the Big Bang didn't really explain the size, shape and uniformity of the bit of universe that we see today. To get the equations to fit what was actually observed there had to be another fudge, and indeed in the early days a fudge was exactly how it was regarded. Somehow this fudge has become part of the foundations of current cosmological theory to the point where it is rarely remembered as being a fudge at all.

The idea was called 'inflation'. This required that, rather than the universe expanding at the rate we appear to see now, size must have increased much more explosively in earlier times. Size doubling rates in some mysterious way depended on the age of the universe.

Just to show the magnitude of this fudge factor, if the universe expanded at the rate we see today its size would have increased eightfold without inflation during those times. With the inflation it would have increased a trillion, trillion, trillion times [2].

Some fudge factor!

These newly 'tweaked' equations now provided the higher degree of uniformity of energy observed in the universe. Even now they don't provide a reason for this rapid expansion. As with much of modern physics, the equations fit the apparent observations but often don't bring with them a conceptual framework to explain why, just as with Newton's spooky 'action at a distance'.

Even Einstein wasn't averse to putting fudge factors in when it came to the evolution and fate of the universe. Indeed, it led him to what he thought was the biggest mistake of his career.

The equations of his General Theory of Relativity indicated an unstable universe, one result of which would be a Big Crunch in which the universe would contract back into one big ball of matter. All life in the universe would be destroyed and a huge black hole would be created. The whole universe would cease to exist.

That was bad enough but the alternative was perhaps worse. This would mean an ever expanding and cooling universe. Again, all life would end as the stars finally burned out. This is because present theories of sun formation indicate that our sun is in fact a third-generation star. In other words, it is made up of matter from the deaths of two previous generations of stars. Mass, scattered after their explosive deaths, comes back together due to gravitational attraction and gives birth to a new star. In an ever-expanding universe that was getting less and less dense, and all stars further and further apart, then this regenerative mechanism wouldn't work and all the lights would go out. The universe would end in a whimper.

Not happy with either grim fate, and inculcated in the thousand-year old view that the universe was everlasting, Einstein put in a fudge factor to his equations, called the Cosmological Constant. This implied an unknown force acting against gravity and balancing it out so that the universe could go on forever. Einstein himself said that '... this exposes me to the danger of being confined to a madhouse.'

Hubble's work, both on the size and expansion of the universe, apparently overturned the idea of an everlasting universe, and this was what made Einstein consider his work such a mistake.

Even if we choose to ignore the Big Bang's manifest failings, the theory still doesn't help explain where everything that makes up the universe actually comes from. However, there were experimental anomalies which led some physicists towards an explanation. Although the spread of galaxies in the universe is fairly uniform, it isn't as perfectly uniform as would be expected from the Big Bang theory.

Some cosmologists grasped at such anomalies to dream up the concept of vacuum fluctuations.

The idea is that the vacuum, rather than being full of absolutely nothing, is actually seething with 'virtual' particles popping into and out of existence.

This idea, and the belief that infinity can be a real thing, led some physicists to come to the conclusion that the vacuum is filled with infinite energy. This vast energy comes and goes but somehow almost always manages to exactly balance out to zero over time.

But at the time of the Big Bang it somehow didn't balance. There was a fluctuation in the 'zero-point energy' and the whole universe was effectively burped into existence.

The Heretic would say this is the latest, and perhaps greatest, of present scientific myths: the idea that the vast universe comes from absolutely nothing.

All the way back to the Greek philosopher Thales in 550 BC there has been a belief in the underlying conservation of some fundamental substance in any process or reaction. Later on, the scientists identified this basic substance as matter and energy and the Laws of Conservation of Mass and Energy were formulated. These Laws were considered immutable and sacrosanct right up until physicists began to believe the universe was burped into existence and then suddenly these venerable Laws weren't so immutable and sacrosanct at all.

Everything from nothing

Let's look at this concept a little more closely so we can appreciate its true splendour.

It comes from a theory supposedly governing the very smallest things, applied without apparent reservation to the largest thing of all, namely the universe. It is called Heisenberg's Uncertainty Principle and can appear at first sight as of little importance.

Essentially the Principle dictates that the more accurately we know a particle's position the less well we know its velocity and vice-versa. It isn't intuitively obvious as to why this is the case but it is something that emerges from the mathematics.

Not convinced? Well the next step gets worse: infinitely worse. Let's assume the particle we are interested in is constrained within a small box. We will know the position of the particle in it to within a small error, in other words the volume of the box itself. According to Heisenberg there will be some uncertainty about the particle's velocity and hence its energy. The smaller the box, and the more accurately we know the particle's position, the less certain we are about its energy.

Physicists even apply this reasoning to vacuum and then take another huge leap. They say that the smaller the volume of vacuum we consider, the greater the energy uncertainty and fluctuations that can occur. If we go to zero volume, the fluctuations can then supposedly be infinite.

And that, believe it or not, is how physicists come to the concept of infinite zero-point energy. In other words, empty space is supposedly boiling with infinite energies which come and go like the breeze.

That infinity may be an artefact arising from mathematics that itself may have no direct coupling with reality, is a thought that troubles few physicists. If it is an artefact this makes the whole idea of zero-point energy nonsense.

But there's still worse to come. As energy and mass are related, according to Einstein's famous $E=mc2$ equation, this would mean that massive energy fluctuations could cause huge numbers of particles to pop in and out of existence. Because of the implicit belief in the mathematical concept of infinity as having some basis in reality, this actually means that at every point in space an infinite number of particles are popping into and out of existence all the time.

And you may have thought the idea of ghosts was crazy!

Though our Heretic is spitting nails at this point the Sceptic has an ace up his sleeve. He believes there is evidence to support this bizarre theory. It's called the Casimir Effect.

In 1948 Casimir and Polder were experimentally examining the forces between atoms and they found that the results didn't quite match what was predicted, although they were out by only a tiny amount. The Casimir force has been measured as about a millionth of a gram of weight. In search of an explanation, they considered the effect being due to these supposed vacuum fluctuations.

The Heretic is having none of this, however. OK, he says, so there's a tiny discrepancy between what was expected and what was measured. Would that be enough for a reasonable person to take it as proof of infinite zero-point energies and a whole universe popping into existence. Or might such a reasonable person believe it more likely this was some other minor but as yet unidentified effect arising from our incomplete knowledge of the various forces between atoms. Some more details of these six other forces can be found in the next chapter for those who really want it?

In modern theoretical physics, where common sense was abandoned about a hundred years ago, the idea of infinite energies and universes appearing magically from empty magicians' hats won the day.

Before we leave the issue of vacuum fluctuations it is worth mentioning quantum mechanics which will be explored in the next chapter. Quantum mechanics is the study of the very small, just as cosmology is the study of the very large. Physicists believe both must be linked by what they call the Unified Field Theory. Many physicists have spent their careers, and Einstein the latter part of his, trying to develop such a theory which would explain everything. Nobody has yet succeeded.

Indeed, the physics of the very large and the very small seems to be entirely different and this is nicely illustrated by the effects of these so-called hypothetical vacuum fluctuations.

These fluctuations mean that short lived particles will pop into and out of existence. While they are 'in existence' they will exert a gravitational force. Weak though this force is, there are supposedly so many of these particles that they should have a marked effect on the fate of the universe.

Unfortunately, the universe shows no evidence of these predicted effects. Indeed, if there is an effect, it will be 120 orders of magnitude less than predicted. In other words, it is in error by a factor of 1 followed by 120 zeros [2].

I have struggled to find an everyday analogy which might give the reader some idea of just how mind-blowingly large an error this is but I can't. For example, suppose you used one technique to measure the distance between the tines of a kitchen fork which should be a distance of a couple of millimetres. However, your measurement erroneously gives a distance equivalent to the 100,000-light year diameter of the Milky Way galaxy. Unbelievably huge though that error would be, it would still only be about 1 followed by 21 zeros. The errors between reconciling the two 'hard' science subjects of cosmology and quantum mechanics, would still be one followed by nearly another 100 zeros greater than that.

Any 'theory of everything' unifying the physics of the very large and the very small, supposedly just around the corner for nearly a century, would seem to actually have a very, very long way to go!

If the truth lay anywhere close to the quantum mechanical view then galaxies would never have formed and we wouldn't be here to make our observations. So big is this difference that some modern-day physicists, showing remarkable faith in their equations, say that the only explanation is that in fact there must be countless universes. We just happen to be in a very unlikely one where this vacuum energy effect is incredibly small. Rather than abandon their theory they would imagine countless other universes into being without a shred of supporting evidence.

The Heretic regards these ideas with utter disdain, and even the Sceptic may shuffle his feet uncomfortably.

Meanwhile the wide-eyed Believer is booking his ticket on Virgin Wormhole to the universe next door.

Dark Energy, Dark Matter

As mentioned in the Introduction, for most of the latter half of the twentieth century, a great deal of effort was spent on deciding whether there would be a Big Crunch. In other words, was there enough mass in the universe for the gravitational attraction to halt the expansion of the Big Bang and indeed reverse it? If so, the universe would then collapse down to the ylem, perhaps for the process to begin again.

Or, if there was not enough matter for this to happen, would the universe continue expanding but getting forever cooler and cooler until all life in the universe perished?

Neither was exactly a cheery outcome but the question as to which road the universe would take consumed vast amounts of research resources. Conflicting experimental findings meant the argument went back and forth, as scientific arguments often do.

However, for over eighty years there had been hints from observation that all was not well with the conventional view of matter and the universe. Back in 1933 a Swiss astronomer called Fritz Zwicky found that clusters of galaxies were orbiting each other faster than their estimated mass should allow. Obviously, such findings depended on assumptions such as that the galactic mass was known and that the theory of red shifting was correct. For the sake of argument, let's assume they were.

The Dutch astronomer Jan Oort later found that a similar discrepancy apparently governing stars in our own galaxy.

Much of these findings could be put down to instrumental inaccuracies, a common fate for data that 'doesn't fit'. However later, and less easily dismissed observational results by the American astronomer Vera Rubin around 1970, corroborated these earlier studies.

Cosmologists needed a new concept to fit their theory to the experimental results, or another whopping fudge factor as our Heretic might regard it. This new concept was of Dark Matter. This is supposedly matter which does not absorb or emit light, as all other matter does, but it does have a gravitational effect. 'Dark' is therefore actually a misnomer, as dark materials absorb light which Dark Matter supposedly doesn't. Rather it is 'Transparent Matter' or 'Invisible Matter', but the name 'Dark Matter' seems to have stuck.

And there has to be an awful lot of this material to make the theory match observation. Indeed, the latest estimates indicate the theory needs there to be about six times as much dark matter as 'ordinary' matter in the universe.

So, the universe has six times the mass we originally thought. Not because we actually saw some new stuff but because our theory needed it. Again, in order not to mess up the theory, this new stuff was required to be invisible.

But it gets much worse. Surely all this extra matter would necessarily mean that the Big Crunch/Dying Whimper controversy was at an end? With all this extra Dark Mass, and assuming it didn't just exist in the minds of a few cosmologists, then surely there would be enough to stop the expansion of the universe and bring it all back to the primitive ylem? The Big Crunch it is then!

Err...apparently not. Because, yet again, the universe doesn't appear to be playing ball with our theories.

In the 1990s, studies of supernovae by two independent teams appeared to show that the universe's expansion wasn't slowing down, as would be the case in both the Big Crunch and the Dying Whimper outcomes of the Big

Bang, but in fact was accelerating. If you believed in the Big Bang then this was a crazy finding. The Big Bang would have caused matter to fly outward but, after the initial explosive impulse, there would be no further acceleration. Gravitational effects in fact would make this matter slow down over time.

Incidentally, these new calculations were based on assumptions that supernovae produce predictable emissions of radiation which may or may not be true, though is widely accepted in cosmology. Basically, these supernovae were dimmer than expected, and this could be explained if the universe was actually accelerating outward.

Even if all these assumptions were correct, then what on Earth could cause this continuing acceleration to happen? Another spooky new concept had to be conjured out of the hat: Dark Energy.

And again, this isn't just the odd bit of extra energy here and there. In fact, if we look at the entire energy of the universe and assume, using Einstein's famous equation, that matter has an energy equivalent, the total energy of the universe is made up of only 4% normal matter and normal energy, 23% dark matter and 73% dark energy.

So, this new stuff that has been made up to fit the equations amounts to over nineteen times the amount of normal energy and normal matter that we know about.

We cannot see or directly detect Dark Energy or Dark Matter. So, in other words, 96% of the universe is invisible to us. We can't see it, we can't feel it but, in order to make the equations fit, the physicists tell us it must be out there.

Even our Sceptic is unsettled by the huge, new and undetectable things that need to be dreamt up but he still has faith. He would perhaps shrug his shoulders and say that when it comes to the unknown then such uncertainty comes with the territory. All the objects in cosmology are extremely far away so it's amazing we can determine anything about them at all. We have discovered so much and we will continue to discover so much more. Eventually we will understand the whole picture and be able to produce a Unified Field Theory which will explain all the matter and energy and forces in nature. All the changes necessary in theory over the last hundred years suggest we may have a long way still to go but he is sure that one day we will get there.

Our Heretic, however, would regard the massive fudge factors we have described in this chapter as evidence that something is even more profoundly wrong with our theories and laws of physics. That in fact they are all constructs purely of our own human making and do not govern how the universe operates. Conjuring all this Dark Energy and Dark Matter and putting in vast fudge factors to explain the evolution of the universe, is simply a sign of desperation. The more closely we measure the universe, the greater the disparities we find between theory and fact and the weirder the theoretical phantasms we need to construct to explain them away.

Summary:

Barely one hundred years ago, we lived in an ageless universe which was about 300,000 light years in diameter. There was matter and there was energy. Objects were gravitationally attracted to each other because...well...because they just were.

However, just a few years later we found ourselves in an expanding universe which was about 7 billion years old and the reason objects moved towards each other was because space itself was curved by the presence of mass.

A few more years on and the universe apparently has grown to, at least, 46 billion light years across and its age has increased to nearly 14 billion years. There would either be enough mass to reel the expanding universe back into a Big Crunch, or there wouldn't be and it would expand forever, getting less and less dense until everything wound down and the lights finally went out.

Then all change again and we are suddenly not only in an expanding universe but it's one that is somehow actually accelerating in its expansion. There is also, appearing out of a conceptual void just like a universe supposedly burped into existence, nineteen times as much matter and energy in the universe as we thought. Oh yes...and we can't see any of it.

There are people alive who have lived through this entire evolution of cosmological theory.

And the strange thing is that at any point in the last hundred years, the vast majority of cosmologists would have said that we had got it more or less right, and that the prevailing theory at the time just needed a few more tweaks to really nail it down. The Believer believes that now.

However, the Sceptic will have long since realised that whatever the prevailing cosmological theory, it is almost certainly untrue. And, if massive factors such as 'inflation' and 'Dark Matter' and 'Dark Energy' are having to be introduced, factors which are not just a few percent out but can be many orders of magnitude different, then this means we are a very long way from the truth. But he still believes that, one day, we will get there.

Our Heretic, however, goes much further than this and says that we are incapable of comprehending the universe, of developing a 'theory of everything'. There are two reasons for this: the nature of our flesh and blood brains which constrain us to think in specific ways, and/or that there are no fundamental laws underlying the universe to begin with. The universe just is the way it is.

Any theories we produce are actually myths, analogies which give us comfort and the illusion we understand more than we actually do.

That there are regularities in the behaviour of the universe, at least in this time and in this place, there can be no denying. Celestial mechanics reveal a high order of regularity in terms of things like planetary orbits. The Sceptic takes great comfort from this fact. To him or her it's evidence of an underlying mechanism however complex.

Whether these regularities are constant or even the same all across the universe is anyone's guess, though it is presently an article of faith held by physicists. If in fact there isn't such consistency then maybe we wouldn't have to resort to fabulous beasts like Dark Energy and Dark Matter to make our equations work. Physics, however, would get even more complex and may not even be up to the challenge at all.

In this chapter we've looked at the physics of the very large and the convolutions and bizarre concepts that have to be introduced by physicists to try to make sense of their observations. If you thought this was unsettling, I'm afraid it gets worse. In the next chapter we'll look at the physics of the very small where the theories just get wilder and wilder.

3

Science of the Very Small

Science has 'explained' nothing; the more we know the more fantastic the world becomes and the profounder the surrounding darkness.

Aldous Huxley

Wyszkowski's Second Law: Anything can be made to work if you fiddle with it long enough.

In the previous chapter on the physics of the whole universe we saw how our understanding of the very large may at least be very limited and that physicists have had to come up with some quite bizarre concepts just to shore up their existing theories. But what about the very small; how well do we understand that?

Our Believer has had little to say so far other than demonstrating unquestioning acceptance of received scientific wisdom but now he shows some animation. He claims that the physics of the very small are a refutation of the views of both our Sceptic and our Heretic. He claims high degrees of predictive accuracy for one of these theories, the Standard Model of particle physics, which predicts the existence of all the particles we can observe and is claimed to be accurate within half of one percent [1].

Even more extravagant claims of accuracy are made for the theory of Quantum Electrodynamics which is said to be accurate to 0.00000001 percent (1 in 10 billion). This has led to the claim that it is the most accurate agreement between theory and experiment for any scientific model tested [2].

We will look at some of these claims and again show that these supposed accuracies are achieved only by the inclusion of a multitude of often enormous fudge factors ('fine tuning' or 'renormalisation' as some of these processes are euphemistically known in the physics community) and by the conjuring up of even more mythical beasts never observed or detected in any form. At least another eleven spatial dimensions, anyone?

To try to understand this means we must go on a brief tour of atomic and particle physics to see what experiment actually shows us and to appreciate the theories that cover it, like a blanket pulled over a bed full of rocks.

The atom

The idea that the world is made of indivisible particles has been around since Greek times but it was only at the turn of the 20th Century that we started to develop the technology to really begin to study these 'atoms'.

The thousands of years old belief that atoms were solid little indivisible balls had to be reassessed after the experiments of Ernest Rutherford in the early 1900s. These involved firing tiny positively charged particles at a piece of gold foil. From classical mechanical considerations, these lighter particles should have bounced off the gold atoms in a manner that would be reasonably predictable, assuming the diameters of the supposedly 'solid' atoms were known.

But the patterns of the scattered particles didn't conform to the theory, leading Rutherford to believe that most of the atom was actually made up of empty space with almost all of the mass concentrated in a central 'nucleus'. Some of the particles fired at the target came right back out again. Rutherford likened these surprising results to firing a 15-inch artillery shell at a piece of tissue paper and having it bounce back at you.

Being on such a small scale, we can never actually 'see' the inner structure of an intact atom. We can only glean information about it indirectly by bombarding it with various particles.

Rutherford's conclusions were therefore based on inference from indirect measurements. Many thousands of experiments using such particle interactions have been undertaken since Rutherford's time without apparently contradicting his idea of almost all the mass being concentrated in the nucleus. That said, almost all were designed and set up with Rutherford's concept as a given. As we will see in Chapter 8, such experiments are complex and are designed to observe very specific things, often to the exclusion of other observations which may contradict the theory.

Nevertheless, in the face of such evidence it would be difficult to refute this theory but one must bear in mind that it is still only a theory. The artillery shell comparison again hints at our dependence on analogy that, when it comes to the complexity of the universe, may hide more than it reveals.

Further experiment led to the conclusion that the nucleus was positively charged and contained sub-atomic particles called protons with positive charges and neutrons that had no charge. This nucleus was surrounded by a shell of rapidly orbiting negatively charged electrons analogous, it is often said, to the way the Sun is orbited by the planets.

This view, however, led to other problems. For example, if some particles in the nucleus are positively electrically charged then why don't they repel each other and fly apart as is the case with charged objects on our everyday human scale?

In order to explain this, physicists were led to the idea of a hitherto unknown and unexpected 'strong' force which bound the particles in the nucleus together and overcame the force of repulsion.

Interestingly, this force supposedly works in a completely opposite way to the gravitational and electromagnetic forces we deal with in the larger world. These latter forces decrease with distance. In order to get the Strong Force theory to reflect the experimental data, this new force had to get stronger the further apart the particles were. In an attempt to find a comprehensible real-world analogy to explain this, the most commonly adopted is that of an elastic band. The further out it is stretched, the stronger the force trying to snap it back.

Examinations of the electromagnetic radiation emitted by heated elements also suggested that rather than having a wide range of orbits round the nucleus, the electrons seemed confined to certain orbits (they were said to be 'quantised' in that they were restricted to specific values). If enough energy was given to an electron in the atom, it would transit to a higher orbit (or shell) and then, if it gave the energy up, it would drop down again to a lower shell.

Real world analogies were becoming difficult to find that might help to understand this process.

And still the theory wasn't explaining the thing that had got people interested in the atomic domain in the first place, namely the radioactive decay of certain atoms. In this, a neutron in the nucleus is believed to decay into a beta particle (essentially an electron) as well as a proton plus another uncharged particle later called a neutrino.

Electrical charge is conserved (doesn't change) in this process as the proton has a positive charge and the beta particle has a negative charge. Thus, a neutrally charged particle produces a positive and a negatively charged particle which cancel each other out in terms of overall charge. The principle of the Conservation of Charge remains intact. This principle has it that charge cannot be created or destroyed. Whilst cosmologists ditched the principles of the Conservation of Mass and Energy when it didn't suit them, the Conservation of Charge is something the particle physicists have struggled to keep.

Incidentally, the argument the cosmologists come up with for rudely jettisoning the long-accepted principles of Conservation of Mass and Energy shows how far they will go. It says that the whole mass of the Universe can be temporarily created by the so-called vacuum fluctuation, but that it will also be destroyed at the Big Crunch. That may happen zillions of years later but, in the end, it will all even out.

For that argument to hold, the billions of years of universal existence is itself just a short-term fluctuation, which is surely stretching it a bit. It also neglects the accelerating effects of that pesky new Dark Energy that had to be dreamed up. That indicates there will be no Big Crunch and so no balancing annihilation of mass or energy, and so no conservation of either.

It appears that physicists, like so many of us, just cannot stick to their principles. Whether Conservation of Charge as an inviolable principle makes it into the second half of this century remains to be seen.

Even the newly minted strong force and the electromagnetic force still weren't enough to explain the experimental findings. On top of that the other known force, gravity, was far too weak to impact on these decay processes. Thus, another new force had to be thought up to make the equations work and this was called the Weak Force, as it was weaker than the Strong Force introduced to explain how the nucleus was held together.

So, in the space of a few years, two entirely new forces came into existence in the minds of physicists though they, of course, would argue these forces had been around since nearly the Big Bang itself.

Protons, neutrons and electrons: physicists were happy with the idea that atoms, before they decayed, consisted of just these three particles. Again and again, this sort of false certainty inevitably leads a respected figure of the time to say foolish things. In this case it led the accomplished physicist George Gamow to conclude in 1947 that: 'Thus it seems we have hit the bottom in our search for the basic elements of which matter is formed' [3].

And this comforting idea of limited complexity in an otherwise fearsomely complex universe held for quite a while longer until bigger and bigger particle accelerators were built. These accelerated particles to higher and higher energies and then smashed them into atoms or other particles. The targets were broken up by the impacts and produced many new particles that didn't have the same behaviour as any of these simple standards.

Mankind has an innate desire to classify and categorise. This is called taxonomy. In biology it allows us to divide all the animals in the world into individual species and sub-species. At its heart taxonomy is trying to make the complex universe simpler and easier to deal with.

So, like Victorian biologists seeking to simplify the myriad creatures of the Earth, physicists set out to classify their newly discovered 'particle zoo'.

Unfortunately, division of a wide range of anything into sometimes arbitrary categories can disguise more than it reveals. The divisions, or categories, are ours not nature's. It can be both a strength but also a severe limitation of all science.

In order to divide the plethora of particles into sub-groups, new properties of these particles had to be thought up and then developed to allow this differentiation. Physicists had a great time thinking up new names for these properties though they bore absolutely no relation to their real-world equivalents. These included spin, charm and strangeness.

But physicists couldn't really assert they had tamed the bestiary of new particles they had created unless there was a story to unite all these new categories of particles.

After many attempts to navigate a way through the particle zoo a theory was produced called the Standard Model which described the universe in terms of four kinds of particle (electrons, neutrinos, up and down quarks) and the two new types of forces (weak and strong). These new quark particles presented a bit of a problem, though. Unfortunately, because of their supposed nature they can never be isolated and so detected as separate entities in themselves because the strong forces between them are so strong. The existence of the quark, so central to this model, can therefore only ever be inferred.

Sort of like Dark Energy and Dark Matter, says our Heretic, and therefore equally unbelievable.

But now, as far as physicists were concerned, atoms had yielded their 'fundamental building blocks of matter' status to protons and neutrons and electrons, and these in turn had been usurped by the quarks which are presently thought to make them all up.

Despite our inability to isolate them, the existence of quarks is considered to have been established by experiments first performed on the Stanford Linear Accelerator. In an experiment analogous to Rutherford's which bombarded atoms with particles, particle physicists bombarded the nucleus with high energy photons. Based on changes in the interaction rate with increasing energy of the photons, it was inferred that the particles in the nucleus themselves had structure.

So, in essence, the 'evidence' for the existence of quarks is an inference based on an inference.

The Standard Model also requires the existence of four particles that are themselves a means of transferring these four forces: gluons for the strong forces, weak gauge bosons for the weak force and photons for the electromagnetic force. Mass itself was to come from the Higgs Boson. The idea behind the Higgs Boson assumes the whole universe is bathed in a so-far undetected Higgs field. Movement of particles through this field induces mass via this boson.

The detection of particles having the predicted properties of the theorised weak force bosons in CERN in the early 1980s has been considered as one of the triumphs of the Standard Model. Recently, claims for the discovery of the

Higgs Boson on the Large Hadron Collider have been made. This was considered sufficient for Peter Higgs, the man who conceived of the boson named after him, to be awarded the Nobel Prize in 2013.

And indeed, albeit with considerable tinkering, this model did predict the results of experiments for years after it was conceived. Or at least no significant experimental results were recognised as contradicting them. This apparent goodness of fit has led the Standard Model to be regarded as one of the most successful theories in science. However, it has major problems which are generally not highlighted when claims for its success are made.

Limitations of the Standard Model

There are a number of problems and limitations that are not addressed by the Standard Model and these suggest it may have only a glancing interaction with the truth.

Firstly, though it too is a force, gravity is not really incorporated into this model. However, as it is much weaker than these other forces, it would not be expected to play a large role in these particle interactions anyway.

A more substantial problem is to do with the so-called hierarchy of particles. The idea behind the Higgs particle may explain why particles have mass but it doesn't correctly describe why they have the masses they do. In fact, the Higgs theory as it stands, combined with our theories of quantum mechanics and special relativity, indicate that these masses should be 10 to the power of sixteen times greater than they actually are. In other words, 10 million billion times greater [1]. Again, this colossal error between theory and experimental observation is one that is rarely mentioned when the supposed success of the Standard Model and Higgs Theory is touted.

Physicists, somehow keeping straight faces, get around this problem by 'fine-tuning' which even Lisa Randall, a noted physicist and scientific communicator [1], describes as an 'enormous fudge'. Possible explanations for why this is needed involve the proposition that there are more spatial dimensions than the three we are used to. There's no evidence that these exist, or multiple universes either, but physicists seem to resort to these at the drop a hat, so certain are they of the sanctity of their equations. They would rather reach for these entirely undetectable things than consider that their theory is deeply flawed.

If you think reaching for a few new dimensions is bad enough, wait until we get onto String Theory where physicists reach for at least another twenty dimensions, and perhaps even more than a hundred.

There's another big problem with the Standard Model. Dark Matter and

Dark Energy also somehow don't make it into this supposedly highly success-ful theory. If the theory was anywhere near correct you would have thought it would have given some indications of the 96% of the universe supposed now to be 'out there' but invisible. Instead of being accurate to 0.5% as claimed, the Standard Model is therefore inaccurate by at least 96%. The Model has nothing to say of what particles make up Dark Matter. Do they have proper-ties like normal matter such as charge, mass, charm or strangeness, whatever they are?

But we still haven't exhausted all the major problems with this Model. There are also other particles, less exotic than Dark Matter, which the Stand-ard Model didn't predict. Additionally, particles seem to come in generations with greater and greater energies (the electron, muon and tau particles are examples of three such generations) and this was also not predicted.

Believing in the Standard Model also requires a major leap of faith. This is the assumption that any momentum not measured in detectable decay prod-ucts is lost to neutrinos. This assumption leads to the conclusion that, from nuclear activity in the sun alone, something like 50 million million neutrinos would need to pass through each of our bodies every second [2]. Has this been experimentally verified?

No.

Supposedly the reason we have difficulty detecting these particles is be-cause, being virtually massless, neutrinos are very difficult to detect. Indeed, attempts to measure them have always produced very much smaller numbers than expected. Nevertheless, physicists cling to this belief because otherwise energy would not be conserved in particle collisions and this whole concep-tual House of Cards would come crashing down. And yes, particle physicists still cling to the Conservation of Energy idea though as a concept it is in-creasingly looking vulnerable as far as cosmologists are concerned.

Interestingly, the Standard Model predicted that the neutrino would be massless whereas subsequent experiments indicated that it actually has some. The Standard Model was extended to explain this but unfortunately this required seven entirely new parameters to be added to make the equations fit. What these parameters mean in any physical sense is not understood.

In fact, the addition of out-of-the-blue parameters to make the equations fit is not new in the history of the development of the Standard Model. The previous version of the Model needed seventeen new parameters [4], none of whose functions are understood, to make the equations fit the data.

This is like fitting a curve to a group of points. No matter how randomly distributed these points may be, you can fit a line connecting them all by adding more and more curves, each with their own curve parameters. The pa-rameters are meaningless (and would be completely different with any other set of points) yet you can use them to fit the points well in that single case.

Perhaps the reader would find it instructive to refer to Wyszkowski's Second Law quoted at the beginning of this chapter.

Is this what is happening with the Standard Model: a desperate attempt to fit the experimental data using parameters which have no real underlying meaning?

A final problem worth mentioning is found hiding in the maths of the Standard Model but it can also be found in String Theory and Quantum Chromodynamics, of which a little more later. At some point in all these theories those pesky infinite quantities arise, usually when a nasty bit of division by zero occurs. The trick physicists use to deal with them is to shuffle the equations round so that, for example, one infinity is divided by another. It is called 'renormalisation'.

In the early days, this way of manipulating infinities to get finite results made many physicists unhappy, most notably Paul Dirac and Freeman Dyson. Even the Nobel Prize-winner Richard Feynman regarded such mathematical legerdemain as showing that theories that resorted to this were intrinsically flawed. *'The shell game that we play...is technically called 'renormalization'. But no matter how clever the word, it is still what I would call a dippy process! Having to resort to such hocus-pocus has prevented us from proving that the theory of quantum electrodynamics is mathematically self-consistent. It's surprising that the theory still hasn't been proved self-consistent one way or the other by now; I suspect that renormalization is not mathematically legitimate.'*[5]

Though these grand old men of physics never came to terms with this process, the younger generation has. Perhaps they are just used to seeing what may be a complete fudge apparently working. But like all fudges, the theory will work for only so long and then it will need even more fudge factors to save it. How many unspecified new parameters will be needed to plug the hole in the Standard Model, or any of the particle theories, next time some uncomfortable new data emerges?

Additionally, there is another innate limitation in all aspects of particle physics in the sense that it takes the reductionist view to the extremes. Reductionism is where science tries to break down complex processes into simpler components. The $9 billion investment in the Large Hadron Collider is really to make things as simple as possible: in other words, to cut out the inherent complexity of the universe and the multitude of interactions between the unique objects that make it up.

This means that particle colliders like the LHC, from which almost all the experimental data of particle physics comes, is about as far from 'real world' as you can get. The LHC is a huge 22km long ring constructed to hold an extraordinarily pure level of vacuum and to operate at near absolute zero temperature. The low temperature is to enable a colossal number of magnets to produce very high magnetic fields. Without these magnetic fields it would not be possible to accelerate, constrain and steer the particle streams.

Even with this high level of control there are still uncertainties which usually make the data less than definitive. Collisions between highly energetic particles produce jets of miscellaneous particles which in turn decay to produce even more particles. The higher the energy, the more particles that are created. New particles produced by new and perhaps unanticipated processes are difficult to distinguish from this multitude.

In order to make any results believable, collisions need to be repeated many times and statistical analysis used. As we will see in later chapters on the 'softer' sciences, the resort to statistics is usually an admission that the real world is intruding into our carefully controlled and reductionist scientific experiments.

Thus, reality smears results even in the tightly controlled LHC. In the face of this statistical uncertainty, hypotheses have to be constructed long before the experiments are run. It is these hypotheses that the statistical methods test for in the data finally obtained. Though statistics has a certain power to show associations it is limited to testing only for specific hypotheses. By starting out with specific hypotheses, and in particle physics they will almost always be based in part on the Standard Model, then there is the danger you won't discover the unexpected because that is not what you will be statistically testing for.

Particle measurements are also, according to conventional quantum mechanical theory, affected by the quantum uncertainty regarding the mass of any particle which decays. Establishing mass therefore will again take additional and multiple experiments and statistical analyses.

So, in particle physics both theory and experiment are prone to considerable uncertainty.

It may also be, as suggested right at the beginning of this book, that perhaps these elementary particles are not identical. After all, nothing else in the universe appears to be identical so why should they be? If millions and millions of particles are produced in these colliders, and they are all different in subtle ways, then perhaps if you look for a particle with very specific properties, such as the Higgs, then you will inevitably find it.

Incidentally, for the Higgs theory to be correct, the Law of the Conservation of Charge would have to be dropped.

Easy come, easy go!

String Theory

String theorists don't make predictions, they make excuses.
Attributed to Richard Feynman [4].

Despite particle physicists making a stout-hearted defence of the Standard Model as being correct within certain energies, it still doesn't include gravity in the theory. When they attempt to include this force, they appear to transcend the scientific method itself or, if you are more cynical, slip the fragile bonds tethering them to any semblance of reality.

Einstein's general relativity has space curved by mass in a smooth and continuous way but this doesn't cut it in the quantum world at all. In fact, if his theory held all the way down to the fundamental particle level, then gravity would become a force of similar magnitude to the others. The trouble is that experiments show that the force of gravity between two electrons is about a million trillion trillion times less than the electromagnetic force between them.

Clearly, we're going to need a really big fudge to explain this. According to physicists it's called quantum gravity and it needs a whole new theory to make it work. By far the most popular of these new theories is String Theory, as measured by total value of research grants given to develop it and the numbers of physicists working on it.

Instead of regarding fundamental particles as particles, adherents of the theory regard them as strings. These strings are about a millionth of a billionth of a billionth of a billionth of a centimetre long so are much smaller than the sub-atomic particles we have been talking about so far.

According to this theory, particles come into existence because of vibrations in these strings. Different frequencies of vibration produce different particles just like different vibrations of a guitar string produce different notes.

Again, we see the power of analogy. It makes everything sound so simple.

This theory, if correct, should provide predictions for new particles produced by string vibrations. Trouble is, to detect these would require particle accelerators with a slightly higher energy than we are able to produce at the moment.

OK, I was kidding you about the 'slightly' bit. In fact, it would take accelerators ten million billion times more powerful than we have at the moment. How long might it take us to develop such accelerators? It's worth noting that it took a quarter of a century to increase the energy of our accelerators by only sevenfold.

In other words, any predictions from String Theory are experimentally unverifiable and probably will be for evermore. And that's where the tether with reality breaks. Up until now, new scientific theories have been considered suspect until they could be experimentally verified. Indeed, unless they made predictions which could later be experimentally checked, they weren't regarded as theories at all, but were simply speculations. Wolfgang Pauli described such theories as 'Not even wrong'.

So, in fact, String Theory is not so much a theory as a hope. The maths is beautiful which leads some physicists to believe it must be true though, of course, equating beauty with truth is simply an act of faith on their part.

It gets worse. String theory is based on there being more spatial dimensions than there is any evidence for. Six or seven more in fact, depending on who you listen to. That's a big improvement, however, as initially the theory indicated there might be 26 such dimensions.

But these dimensions aren't like the ones we experience, in other words the forward-back, right-left and up-down three dimensions that we know exist. These three dimensions we are comfortable with extend in all directions. However, the new proposed extra dimensions are curled up on themselves in tiny little hyperspatial balls (hyperballs perhaps?). The idea is that gravity is so weak it doesn't have much effect on the quantum world because it leaks away from our limited 3-D world into these higher dimensions.

Begging to differ, the Heretic thinks it leaks away to Narnia.

As with all theories, there are different versions of String Theory and one of the most prevalent involves M-space. Nobody is clear what the M stands for, but suggested candidates include membrane, mystery, magic, murky, mother of all, and even masturbation, the latter coined by someone who I'm guessing wasn't much of a supporter. In M-theory the strings aren't simply one dimensional vibrating strings but instead may vibrate in a plane. Hence the term 'membrane', though to sex it up a bit, scientists talk about Branes instead.

But then gravity messes things up again. Combining String Theory and gravity produces estimates of the Cosmological Constant (the factor introduced by Einstein to 'hold back gravity') which are out by a fact of 10 to the power of 56 (that's about a hundred trillion, trillion, trillion, trillion, trillion, trillion, give or take). This is one of the greatest discrepancies between theory and experiment that has ever been achieved [4].

Fudge factor wise, physicists were really going to have to pull out all the stops now if they were to make this whole idea fly.

Making String Theory even remotely realistic has stretched the imagination of the theoretical physicists who believe in it. The theory has to explain all particles, energies and forces but its complexity provides many possible interpretations. In order to explain things there is one faction that, not for the first or the last time in physics, has reached for multiple and perhaps even infinite universes. Strings are supposed to have different properties in different universes.

This leads on to the so-called Anthropic principle that was mentioned in the last chapter. The vast majority of string-based universes would not support life. For example: after the Big Bang, if there was too much energy, the whole universe would fly apart before gravity could make the exploding bits of matter coalesce into suns and planets. Life would never have formed. In fact, the chance of getting all the balances of physical constants and energies right is extremely unlikely amongst all the combinations possible.

The way round this is to come to the conclusion that just by chance we happen to be in just such an unlikely universe. Otherwise life wouldn't have been created and evolved to get to the point to even ask these tricky questions in the first place.

We can't observe these other possible universes in any conceivable manner so, of course, the issue can't be proved one way or the other.

So, in order to keep the wheels on this whole nebulous theory, we've had to resort to multiple new dimensions and even, according to one faction of string theorists, multiple universes. None of these are remotely observable or their existence otherwise provable and that departs entirely from the scientific method where, in order to establish a theory, one has to experimentally verify it.

If String Theory can't in any way be proved, then how is this different from any religious belief, no matter how absurd? Indeed, the whole issue of a theory which is experimentally unverifiable has led many physicists to consider String Theory more as a cult or a faith than a science.

Proponents of String Theory seem to set more store in beauty than experiment. The physicist John Schwarz even went so far as to say: 'string theory was too beautiful a mathematical structure to be completely irrelevant to nature' [6]. This hang-up is part of a wider malaise, as illustrated by this statement from Paul Dirac, another highly respected physicist: 'it is more important to have beauty in one's equations than to have them fit experiment' [7]. Readers are referred back to Chapter 1 and the inherently unrealistic nature of mathematics for reasons as to why this may be entirely inappropriate.

Despite this colossal silliness, String Theory has somehow managed to become the new orthodoxy.

Nevertheless, many physicists (thank goodness!) hate the idea of having to resort to the Anthropic Principle and believe, like Einstein, that one set of laws should explain everything and shouldn't have to rely on zillions of universes. So vitriolic has the debate become that Stanford Professor Leonard Susskind, one of the originators of String Theory, has described it as a 'war'.

[If you've already had your fill of particle physics then perhaps you should skip the next section on Supersymmetry, but it's worth tuning back in for the later section on Quantum Mechanics]

Supersymmetry and other Grand Unified Theories (GUTs)

Physicists for over a hundred years have been convinced of the fundamental importance of symmetry to the Laws of nature. In physics, symmetry is related to the idea of conservation and the laws that supposedly govern them, for example Conservation of Charge and Conservation of Energy. These things are believed to be preserved during particle interactions (except of course when they get really, really inconvenient). For example, Paul Dirac used symmetry considerations to predict the existence of the positron back in 1927, and a particle matching this description was indeed later detected. These considerations of symmetry mean that physical laws are considered to be equally valid whether moving forward or backward in time. As with all physical Laws, they should also apply everywhere in the Universe.

Symmetry is a big deal in modern physics. For example, according to the Standard Model, each of the electromagnetic, weak and strong forces brings with them their own symmetry. Without these symmetries, Special Relativity theory and Quantum Mechanical theory generate nonsensical results, such as the probabilities for high energy interactions coming out as greater than one. In other words, they are more than certain. That, even for physicists who believe in the reality of numbers, is an absurdity.

But there is evidence that symmetry is in fact not universal. Symmetry leads to the expectation from the Standard Model that there should be as much matter as anti-matter in the universe whereas in fact there appears to be hardly any naturally occurring anti-matter.

And it gets worse. The popular Higgs theory, that there exists a field suffusing everything that induces mass in particles by the intermediary effects of the Higgs Boson, requires a breakdown in weak force symmetries. The equations then fit nicely so symmetry, in this instance, is pushed aside.

Weak interactions also violate parity symmetry in that particles 'spinning' (in quantum mechanical terms and not real world spinning top terms) to the left interact differently than particles 'spinning' to the right. In other words, the left-right equivalence breaks down which again is bad news for symmetry.

With all other forces we are aware of, right and left are treated equally but for some reason it doesn't work for the weak force. This led to the concept, nonsensical in real world terms (but then again, with new dimensions and countless new universes being postulated in other areas of physics, should we be surprised?) that the Higgs Field puts infinite weak charge into the vacuum. Infinity, which physicists and mathematicians try to shy away from in their equations, suddenly comes in handy here and instead symmetry is ditched.

Or an alternate argument has it that symmetry can under certain circumstances be 'spontaneously' broken. This means all the other symmetry considerations and Laws of Conservation continue to function to keep the theories governing the rest of physics working, but in this particular instance they somehow don't. The laws of the universe continue to act symmetrically except when they get in the way of our presently favoured theories.

So, such physicists, though not known for being a sporty bunch, appear to have no hesitation in moving the goal posts.

Though the inviolability of symmetry had taken a bit of a beating with weak forces and the Higgs Field, it came roaring back with the theory of Supersymmetry. This is another mathematical framework with supposedly explains the hierarchy problems (three generations of some particles with much greater masses than predicted by the Standard Model). It says that all particles have a supersymmetric partner called, and I'm not making these up: winos, squarks, charginos, and steptons. These interact in the same ways but with different quantum mechanical parameters. A consequence of this theory is that there would be many more types of particle with known masses that should have been detected.

Unfortunately, they haven't been detected, which suggests that the masses of the partners can't be exactly the same as the originals. That in turn means that even Supersymmetry needs to be broken to make theory fit with experiment.

So, is the universe symmetrical or not?

In an attempt to explain this inconsistency, Lisa Randall has written: 'The best scientific descriptions frequently report the elegance of symmetric theories while incorporating the symmetry breaking necessary to make predictions about the world'. She has also likened symmetry to an email spam filter which, having separated the wheat from the chaff, can be discarded.

Incidentally, whilst the Standard Model needs seventeen unexplained and undefined new parameters to balance the books, Supersymmetry needs 105 more.

Let's just quote Wyszkowski's Second Law again: *Anything can be made to work if you fiddle with it long enough.*

There are other alternative theories to Supersymmetry to explain the hierarchy problem. One is called technicolour which requires an entirely new force and, of course, at least one new dimension.

There is no evidence for either and, as life is short, let's not go there.

Quantum mechanics

But even with all these complex theories and their plethora of new parameters introduced to make the equations fit the observed data, things were still not quite right in the world of physics. The old analogy of an atom being like the Solar System, with the electrons orbiting the nucleus like planets round a sun, just really wasn't explaining the strange behaviour of electrons. In some observations, the electrons behaved in some ways as though they were particles and in others they behaved as if they were waves.

Whether things were particles or waves has been a controversial issue for much of recorded history. The particle/wave nature of light has been the subject of conjecture stretching all the way back to the Ancient Greeks and this has swung back and forth for over two thousand years. Aristotle thought light was in the form of disturbances in one of the four fundamental elements (earth, air, fire, water), whereas Democritus was of the opinion that everything in the universe was made of particles that could not be broken down into smaller units. This latter view represented the first of the Atomist approaches. Alhazen, an Arabic scientist in the 11th century, in the first significant treatment of the principles of optics, used atoms as his starting point. Rene Descartes, on the other hand, in the 17th century considered light from a wave-like perspective. Later in that century Newton, however, was very much of the opinion that light was corpuscular, in other words particle-like, though contemporaries such as Hooke and Fresnel disagreed.

Things seemed to settle down when James Clerk Maxwell came up with his theory of electromagnetism in the following century which best fitted experimental results and which definitely had light in wave form.

All was well until the turn of the 20th Century when Max Planck developed a theory that again suggested the particle-like behaviour of light. Einstein seemed to follow suit with the work which won him the Nobel Prize that explained the photoelectric effect. In this, ultraviolet light shining on some metals cause electrons to be emitted. Einstein's theory explained this effect in terms of light being in the form of discrete packets of energy. These packets eventually came to have the name of photons.

Unfortunately for Einstein, his theory involved the quantisation (setting to certain discrete numerical values) of energy, at least in terms of electron shell transitions. That's like pulling onto a motorway and expecting cars to have a wide range of speeds, but instead finding cars travelling at exactly 40mph, or 50mph or 60mph. This had ramifications that Einstein was distinctly ill at ease with and fought against for the rest of his life.

It's worth going into some detail about this issue of whether light is made of waves or particles. It caused a meltdown in physics and has produced theories which either seem to put men back at the centre of the universe, just like in pre-Copernican days, or have them reaching for an infinite number of universes (now, where have we heard that one before?).

Back in the late 18th Century, Thomas Young had shown that by passing light through double slits the pattern that appeared on the screen behind the slits, rather than showing two bright narrow bands, showed a range of light and dark bands. This seemed to indicate that light emerging from each slit acted as a wave. These two waves then interfered with each other to form bands of constructive (additive) interference and black bands where the interference was destructive (subtractive) and therefore cancelled each other out.

However, as later experiments on lasers showed, each impact of light on the screen was essentially particle like. In other words, the light behaved as particles rather than waves.

These apparently contradictory results eventually led onto the concept of the wave-particle duality of light, in other words light sort of behaved as both. This idea was later extended to the wave-particle duality of all matter. So, for example, electrons in similar situations can be seen to behave as individual particles on some occasions or as waves in others.

Now one might suppose, based on the to and fro nature of the debate that has been going on about this issue for two thousand years, that rather than behaving like particles or waves, that perhaps photons are something entirely new that don't have a corresponding analogy that our limited human brains can deal with. In other words, perhaps these things act neither as particles nor waves but something entirely beyond our ken. Our hard-wired thought processes are so hung up on analogy that we can't even begin to understand them. Analogy is useful but can again lead us astray as is illustrated in the next section.

This was all bad enough but then along came an observation that really rattled the cages of the poor old physicists. When the dual slit experiment was allowed to run without close human observation, then wave like effects were found. However, when the individual particles were each observed they acted like particles.

Physicists became obsessed with the idea that the act of observation itself fundamentally altered the behaviour of the photons.

Suddenly man was again central to how the universe worked.

In order to explain these phenomena physicists started to resort to overheated theories far beyond anything dreamed up in science fiction or fantasy literature. Three main competing theories were put forward.

Here they are in descending order of present acceptance by the physics community:

1. The Copenhagen Interpretation and Schrodinger's Cat.

This theory can be summed up by the well-worn tale of Schrodinger's cat.

Don't worry too much as this is an entirely imaginary 'thought experiment' and no actual cats have ever been harmed in its making.

Let's say, however, that some poor cat is trapped in a box whose contents cannot in any way be humanly observed until the box lid is lifted. There is a phial of deadly gas in the box with the cat. The gas will be released if a particle detector, also in the box, detects emissions from a radioactive source next to it in a certain time interval. Whether or not the source produces such an emission in this time is random, or at least that's the present belief. If a particle is detected, the phial is opened and the cat dies. If it isn't detected, the cat lives. The question Schrodinger raised is: after the time interval, but before the box is opened and the health of the cat is actually observed by the experimenter, is the cat dead or alive?

In the real world we'd say it could be either but in the quantum world of the Copenhagen Interpretation the cat is both alive and dead simultaneously. The concept is that until the box is opened the cat simultaneously inhabits two states, one in which it is dead and one in which it is alive. Only when the box is opened and a human observer looks at the cat does he or she see only one state. In physics terms, the act of observation by a human being collapses the wave function and constrains it to only one state.

Though the cat/poison gas/radiation detector scenario is hardly an everyday experience, the whole idea of simultaneous states has unbelievably huge consequences. That's because countless quantum events occur in everything everywhere in the universe every microsecond. This means that everything is in a whole series of undetermined alternative quantum states until, supposedly, someone comes along to observe it.

If this sounds crazy to you then you're in good company. Einstein once said: 'Do you really think the moon isn't there if you aren't looking at it?'[8].

So, when a hurricane hits a forest, trees aren't actually knocked over. Instead these trees are both standing and knocked over at the same only. Only when someone, you for example, strolls by and looks at the scene do the trees actually assume the state of either being knocked down or still standing.

What about if a cow was wondering through the forest instead of a human? Would the cow also 'collapse the wave function' so some trees assumed the knocked down state? That's a tough question to which even adherents of the theory can give no convincing answer.

But, according to this viewpoint, at least man (and maybe the odd cow) is again at the centre of the universe. Things can only take one form or the other when he or she looks at them. Until then they exist in simultaneous superimposed ghost-like alternative states.

This is all very weird and it's hard to believe that physicists could come up with an alternative theory anywhere near as crazy.

But then they did.

2. Everett's Many Worlds Interpretation.

In this theory every single quantum event shears off two entire universes: the only difference between the universes being in the results of that quantum event. So, to make use of poor old Schrodinger's cat again, in one universe it would be dead, in another it would still be alive.

The trouble is that every single quantum event, countless numbers of which are happening in each tiny part of the universe every tiny fraction of a second, will each generate two new universes. So countless numbers of entire universes will be being generated every second. These universes, as far as we know, are unobservable so there is no way the theory can be proved.

At least in this interpretation the observer is no longer central to how reality proceeds. That's sort of a good thing as the last thing humans need is to be even more arrogant about their place in the universe than they already are.

3. Hidden Variables Theory

Basically, this says that the theories of quantum mechanics are incorrect and that quantum effects aren't probabilistic. Einstein was a proponent of this view and said, with others, that 'elements of reality' which were yet to be discovered would need to be added to show that events weren't probabilistic but instead determined. These elements came to be known as hidden variables.

This theory is not highly regarded today as it brings with it the possibility that these hidden variables are a sign of action at a distance with faster than light transmission of information. It was to counter these sorts of ideas that Einstein looked for an alternative to Newton's theory of gravity in the first place when he produced his General Relativity theory.

Perhaps one other reason Hidden Variables theory is in disfavour is that it inherently suggests there are fundamental flaws in our understanding of physics. Some physicists would resort to countless new universes springing into existence out of nowhere than consider that possibility.

Needless to say, many physicists were unhappy with the Copenhagen interpretation. Einstein was one of them: 'Quantum mechanics is very worthy of regard. But an inner voice tells me that this is not yet the right track. The theory yields much, but it hardly brings us closer to the Old One's secrets. I, in any case, am convinced that He does not play dice›.

Such physicists took the view that quantum mechanics was essentially incorrect and that quantum events were not probabilistic at all. That would mean, for example, that radioactive events weren't random at all, as is conventionally considered to be the case. It suggests instead that we were simply unaware of the underlying factors that precipitate each radioactive event.

The Hidden Variables approach is sort of suggesting that somehow each particle 'knows' whether it is being observed or not and acts accordingly. That could explain the dual slit results where sometimes particles acted as particles and sometimes as waves.

Originally it was envisaged that each particle had a 'pilot wave' associated with it that went through both slits and helped steer each particle through one slit or the other. However, the act of observation interfered with this pilot wave and destroyed the interference pattern. This was the idea of the physicists de Broglie and Bohm [9].

The trouble with this idea is that it assumes what is called non-locality, in other words it requires instantaneous (and thus faster than light) communication. Einstein, echoing his original departure from Newtonian mechanics, said we were back to 'spooky action at a distance'.

If you take the Hidden Variables idea a little farther then you are encroaching onto realms of philosophies like Buddhism. Each individual particle would seem to someone 'know' if they are being observed by any one of seven billion human beings on the planet. That smacks of the idea that each particle may be essentially omniscient.

So, in summary, this single anomalous wave/particle observation has prompted respectable scientists to suggest that the observations each of us makes, however humble a person we may be, constrains a whole universe to a single existence (state). Or, alternatively, that each tiny event spawns a new universe. Or that each little particle knows far more about the universe than we do.

That physicists have to resort to such explanations is used as an argument that human 'common sense' is limited and faulty as none of these make common sense at all. However, one would argue that to resort to three such radically different explanations is itself pretty absurd and maybe we shouldn't be abandoning using common sense quite so quickly.

Whatever explanation you choose has profound philosophical interpretations but is it possible they are all simply a consequence of our unreal mathematics? Or could it be that our innate thinking based on analogy has got us hung-up on whether light is analogous to particles or to waves. Perhaps it is not even close to either but, hidebound by our need for analogy, we can't even begin to comprehend it.

You'll probably be relieved to hear we're nearly at the end of the science of the very small but there's one more thing we've got to deal with as our Believer thinks another theory has such a high degree of accuracy so it must be true. If we're going to show the fallaciousness of this then I'm afraid it's back to those damned vacuum fluctuations. If you're feeling faint and can live without this then please jump to the last section of this chapter.

Quantum Electrodynamics and Virtual particles

The supposed wave-particle duality leads to uncertainty when it comes to estimating both position and momentum according, at least, to Heisenberg's Uncertainty Principle. As has been mentioned before, this concept leaks over in the equations to similar considerations for energy and time. The effect of this is that the shorter the time period we observe a volume of space then the less certain we can be about the energy within it. If we only observe for a short period of time then supposedly there can be huge energies within this volume of vacuum.

And energy means matter, according to Einstein. So according to the equations, if not the reality, empty space is therefore bubbling with particles coming into and then out of existence. Perhaps struggling with the lexicon to find a word which didn't make this concept sound silly, the name 'virtual' has been applied to these particles.

Virtual particles come into existence because of 'fluctuations in the vacuum'. Indeed, once some credence was given to this concept, other physicists working on the opposite size scale, the cosmological, came up with the idea that the whole universe is a sort of a vacuum fluctuation though somehow sustained over many billions of years. More latterly the supposed existence of Dark Matter and Dark Energy suggests this is not a fluctuation at all. In that case presumably such particles lose their virtue and become real, but we'll ignore that inconvenient point as we try to explain Quantum Electrodynamics.

Even though the virtual particles are popping in and out of existence they can for a brief time interact with 'real' particles. This interaction was incorporated into the theory to describe the relationship between photons and electromagnetic interactions called Quantum Electrodynamics or QED. QED was used to make predictions about specific experiments on charged particles which were supposedly accurate to 0.00000001 percent, though remember that that requires assigning values to seventeen undefined and mysterious new parameters to work. Again, we may start wondering if Wyszkowski's Law is coming into play again. Despite all this, it has been claimed [2] that the theory is the most accurate correspondence between theory and experiment ever tested on Earth.

Sounds good on the face of it but it should be pointed out that QED theory didn't predict the existence of the neutron (experimentally discovered in 1932) or the muon (1937) or the pion (1947). Indeed, QED is a theory which doesn't have any exact solutions for its underpinning equations. The way of calculating its effects requires what is called a perturbation expansion which is like a mathematical power series where you have a series of terms extending to infinity.

The idea behind a power series is that if you only calculate the first term you will only get a rough approximation of the final result. Including the second term will give you a better approximation and so on until, when many higher order terms are included, you get close to the correct solution.

The trouble with QED is that while the first and second terms in the series can give a good approximation to the experimental results, subsequent terms make no sense whatsoever. Rather than each additional term bringing the solution converging on the correct result, even just the third term becomes infinite.

When this was noticed it led many physicists to assume the model was deeply flawed and indeed QED languished for a while. However, 'renormalisation' was introduced. According to Woit [4] 'this involved trying to make sense of something that was infinite by subtracting something else infinite according to somewhat unclear rules'. Mathematicians at the time found these physics calculations 'mystifying and quite alien to their own world', but when did that stop them?

After many years of tinkering with the theory and with the fudge of renormalisation being repeated so often it became accepted without much question, and the seventeen new parameters being quietly assigned though without any conceptual underpinning for any of them, then QED sort of looked like it was getting some things right. Somehow this became considered the most successful theoretical prediction of experiment known to man.

However, when physicists tried to apply this QED approach to studies of the strong interaction and quarks, called Quantum Chromodynamics or QCD, then this high level of accuracy falls away [2].

Further tinkering is clearly required.

Summary

In order to explain the workings of the world of the very small, modern physics has brought us: virtual particles popping in and out of existence; empty space containing infinite energy; extra undetectable dimensions; infinite numbers of undetectable universes; the massive decoupling of the theories of the very small and the very large; theoretical models that require 17, or even 105, undefined numbers to produce predictions that are accurate to 0.5% in some ways but wrong by 96% in others; theories going so far beyond the real world that any predictions they make can never be tested.

One couldn't, as the saying goes, make it up.

Even our Believer must baulk at accepting theories that can never be verified.

The Sceptic says we're just going through a bumpy patch at the moment, that it is a sign that existing dogma is in the process of being superseded. He or she is sure a new Einstein will sooner or later emerge and bring with him or her a profound new insight allowing us to rewrite the laws of physics, to reconcile all these conflicting ideas and findings and make all things clear again. After all, the history of science is of the apple cart of dogma and convention being periodically overturned and this is surely just one more example. Each overturning of dogma is followed by new ideas which bring us closer to the truth. Who knows how many more upsets will be required, how many more geniuses like Einstein and Newton, how many hundreds or thousands of years it will take? It doesn't matter because sooner or later man will understand the workings of the universe.

To the Heretic, however, all these crazy ideas smack of desperation. To him or her the physics is broken, if in fact it ever worked at all. We carefully fit our theories to explain experimental results but these are getting more and more contorted as more and more data comes in. We have to add more and more undefined factors, or universes or dimensions to try to get some correspondence between theory and reality. To the Heretic these theories and equations and laws are entirely of our own making and do not reflect reality. Again, they are simply myths we tell ourselves to make us believe we understand a universe which is not explicable, at least to the human mind steeped as its thinking is in the hard-wired need for analogy.

The Heretic thinks our theories are simply a projection. Resorting inevitably to analogy, he or she sees our physics as again like the lines of latitude and longitude we project onto the surface of the earth. As far as the Earth and its workings are concerned, these lines have no relevance whatsoever to its reality: how it works, how it exists, its complexity, its past and its future. Nevertheless, latitude and longitude are very useful to us and allow us to navigate over it with great success though they are a purely human fiction.

The Heretic has to agree that physics and the other sciences can be useful. Experiment can show regularities in the behaviour of the universe (at least here and at least now) that we can then exploit.

But the theory we subsequently construct to explain our experimental findings may have nothing to do the truth.

4

Black Holes and other Mythical Beasts

I know that most men -- not only those considered clever, but even those who are very clever and capable of understanding most difficult scientific, mathematical, or philosophic, problems -- can seldom discern even the simplest and most obvious truth if it be such as obliges them to admit the falsity of conclusions they have formed, perhaps with much difficulty -- conclusions of which they are proud, which they have taught to others, and on which they have built their lives.

Tolstoy [1].

Let's talk about some aspects of physics which arise from physicists' slavish beliefs in the reality of the purely mathematical concepts of zero and infinity that appear in the equations they use to represent the workings of the universe. Rather than realise these represent instances where the mathematics break down in their representation of the truth, physicists have been led to believe in the absurd anomalies that arise.

Science creates its own myths but then, perhaps invariably, it outgrows them as messy reality intrudes. For thousands of years or more the Earth supposedly swam through something called the aether until Einstein's Special Theory of Relativity seemed to show that it was no longer required for man's explanation of how the universe worked. For hundreds of years there were the Laws of Conservation of Energy and Mass which the latest theories on Dark Matter and Dark Energy suggest do not even begin to apply. The universe was supposedly created in a vast vacuum fluctuation called the Big Bang and the only way mass and energy is going to be conserved, and the equation balance, is if it all comes back together and vanishes in the Big Crunch. Unfortunately, the proposed Dark Energy is supposed to counteract and indeed overcome the effects of gravity. This means the universe is flying apart faster and faster and there will never be a Big Crunch and the equations will never balance. The Law of the Conservation of Charge also seems to have been abandoned in the case of the theory behind the Higgs Boson.

In the world of physics, it appears nothing is sacred. Therefore, let's look at things which are more of less considered gospel right now but which will in future almost certainly be seen as similar myths if history is anything to go by.

Black Holes

The idea of black holes is an old one. As far back as 1783 John Michell at the Royal Society suggested that if a star was sufficiently massive then 'corpuscles' of light would not be able to escape it, rendering it invisible.

Black holes exist, at least according to most books on the subject. Take Prof Lisa Randall of Harvard University, one of Esquire magazines '75 most influential People of the 21st Century' [2]: 'The Black Hole at the centre of our galaxy is 10 trillion metres in radius' and 'Black holes do, however, exist throughout the universe- in fact they seem to sit at the centre of most galaxies'. Not much doubt about their existence there.

Watch popular science programmes on cosmology and Black Holes nowadays seem as established as craters on the Moon.

But the Heretic regards Black Holes as simply a consequence of believing in the reality of the mathematical concept of infinity. It is Einstein's General Theory of Relativity which leads us down this rabbit hole. Black holes supposedly arise from massive stars. During most of their lives these burn by the hydrogen based nuclear fusion process and this produces outward pressure which balances the gravitational effects that would otherwise make the star contract.

Eventually, however, the sun runs out of enough hydrogen to sustain this balance, gravity becomes dominant and the star collapses in on itself. What happens then depends on how much mass the star had to begin with, at least according to current theory. Some don't have much mass and become a very big cinder. But with enough mass the electrons and protons combine to make what's called a neutron star. That's another body springing from theory which has assumed the mantle of being a stone-cold fact. Indeed, so confident are physicists in the existence of neutron stars and the ability of these to capture dangerous new objects such as Black Holes, that they are used to 'establish' the safety of the Large Hadron Collider as we will see in the chapter on risk in science.

If, however, the star is even more massive then, according to the mathematics, it collapses into a nasty zero sized space. And, of course, wherever there's a zero in a mathematical equation there's usually an even nastier infinity lurking nearby. If you have something with very large mass occupying zero space then the equations would have it that the curvature of space-time becomes infinite. In other words that point in space becomes what's called a singularity.

That's the equivalent of tearing a hole in space-time.

So great is the bending of space-time that nothing, not even light, can escape such an object. That obviously will make them pretty difficult to detect. Nevertheless, scientists believe they can infer their presence by their gravitational effects on adjacent stars.

For example, and assuming we really do understand gravity throughout all time and space, measurements of the behaviour of adjacent stars indicate that there is a Black Hole in the direction of the constellation of Sagittarius weighing in at about the equivalent of 2.5 million of our own Suns.

However even the Sceptic would have to admit that we are still in the very early days of understanding the cosmos. That distant stars seem to be accelerating away from each other rather than de-accelerating supposedly due to Dark Energy shows there are many processes we do not understand. In the light of such ignorance about these basic mechanisms it is likely that, sooner or later, this relatively early concept of Black Holes will be abandoned.

Interestingly, even though Einstein dreamed up General Relativity and constructed its equations, he always denied the existence of Black Holes. Some critics argue that, like his unwillingness to believe in the conceptual underpinnings of the quantum world, in its dead cats and multiple universes, that this shows a final failure of nerve and imagination on the part of an undoubted genius.

Maybe, or perhaps he just smelled a rat. He once said: 'As far as the laws of mathematics refer to reality, they are not certain, and as far as they are certain, they do not refer to reality.'[3]

Thousands of years ago, people did something similar when they thought the twinkles in the night sky were the light from the distant campfires of tribes who lived there. In making the connection between some mathematical expressions and the observations of star behaviour, physicists are essentially doing the same thing: trying to project their own limited understanding onto the workings of the universe.

Will the concept of Black Holes stand the test of time? Judging by the staggering upheavals in cosmology over the last hundred years its chances must be low.

Wormholes

The idea of wormholes, those concepts beloved of science fiction that allow faster than light travel between vastly separated parts of the universe, itself springs from a division by zero causing an infinity in the equations governing general relativity. To understand what a wormhole is, you can think of the line in space between two distant stars as a long thin flexible ruler. If space is warped in an Einstein-like way then you can think of this as perhaps bending the ruler until the ends nearly meet. If you could somehow produce an infinity-requiring 'wormhole' in space and time between the ends of the ruler then you'd have much less far to travel to get from one star to the other than if you went along the whole length of the bent ruler.

Of course, there is absolutely no evidence that such things exist. Perhaps we shouldn't hold our breaths in anticipation of visiting the stars any time soon.

Even worse, our Heretic believes we never will (it's no fun being a Heretic). Light travels around thirty thousand times faster than our best current propulsion systems are capable of producing, yet the nearest star is so far away that light takes over four years to get there. Even assuming that one day we can build spacecraft a hundred times faster than they are today, that still means journeys lasting over a thousand years. What device, even of the simplest kind, has worked for over 1000 years? And, of course, a spaceship that can travel between stars would be the complete opposite of simple and would have many systems and component parts that could break down. Or more likely, go wrong in all sorts of unexpected and entertaining ways.

Thus, our killjoy Heretic believes we will never visit the stars and any aliens out there will never visit us for the same reason. Why there aren't aliens here on Earth already (the so-called Fermi Question) is one of the outstanding mysteries discussed in the Appendix, with this and other possible answers to the question being provided.

Singularities and the Rapture of the Nerds

Singularities are where mathematical functions blow up. For example, when you divide one by zero then suddenly you get an infinity rearing its ugly head.

The naive belief in the concrete existence of infinity has spawned an aspect of pseudo-science. This is the creed of the 'Technological Singularity'. The idea of such a singularity was coined in the 1950s by John von Neumann. He considered that the rapidly accelerating progress in technology indicated an approaching singularity in terms of the history and way of life of humans on the planet. So huge would this change be that our technological capability would become infinite.

This concept of a technological singularity was taken up by science fiction writer Vernor Vinge, and by inventor and futurist Ray Kurzweil who said that an intelligence explosion was therefore coming. Essentially computers would themselves start designing better and better computers until extremely powerful minds would be created which would be well beyond the capacities of the human brain to even understand. Vinge and Kuzweil predicted the singularity event (the creation of an artificial super-intelligence) in 2030 and 2045 respectively [4, 5].

In order to lend these ideas some credibility, the analogy that is often used for this exponential growth is Moore's Law. This has predicted, to reasonable accuracy, a doubling of the number of transistors that can be deposited on integrated circuits every 18 months. The more transistors, the greater the computing power.

A super-intelligent machine designing even more super-intelligent machines is seen as producing a rapidly accelerating cascade of intelligence.

But perhaps humans won't be left behind, unless of course the machines wiped us out. Kurzweil thinks that: The Singularity will allow us to transcend these limitations of our biological bodies and brains...There will be no distinction, post-Singularity, between human and machine [4]. Human minds will apparently be uploaded to computers and we will become immortal Gods.

The Heretic and a horde of other non-believers refer to this as the Rapture of the Nerds.

The whole idea has even gone on to spawn the California based 'Singularity University' which was established in 2008. The President is Peter Diamandis, Founder and Chairman of the X PRIZE Foundation, with Kurzweil, who had become Director of Engineering at Google, taking on the chair of 'Future Studies'. The mission statement of the university is 'to assemble, educate and inspire a cadre of leaders who strive to understand and facilitate the development of exponentially advancing technologies in order to address humanity's grand challenges.'

Lest you think that in this section I am taking lazy pot-shots at harmless cranks crying in the wilderness, it is worth pointing out that von Neumann was a world leading mathematician and Kurzweil has a distinguished history as an inventor. Singularity University has been sponsored by Google, Nokia, NASA and Genentech amongst others.

Obviously in this book our Heretic is arguing against many of these ideas. Firstly, if there is no such thing as infinity in reality, then nothing, including technological knowledge and capability, will ever get to that point. Moore's law has applied reasonably well so far but it is certainly not some underlying law of nature and seems hardly likely to extend to infinity. This is not least because there will be physical limits on the size of any components that can be constructed and the ability to cool the heat generating transistors themselves.

Another major problem with the whole idea is that a computer does not think like a human being and almost certainly never will, as we will see in Chapter 7. As we don't even begin to understand how the mind works, or the brain for that matter, how can we ever replicate it on a computer?

And finally, endless improvements in technology and intelligence will only prove ultimately futile if all our science is based on as unrealistic theories as the Heretic would argue. The present regularities of the universe, such as that apples fall from trees, are indeed exploitable but only up to a point. If true, this means science and technology as we know them will eventually run out of steam.

Some evidence for this latter point can perhaps be found in the numbers of scientists working at any given time and the significant discoveries they produce. A few centuries ago the number of scientists in the world could be measured in the hundreds and yet they laid the foundations of physics and chemistry and biology. Nowadays millions of scientists are employed and yet where are the stream of profound discoveries?

The low hanging fruit of science has been well and truly plucked. There will be more to come, but they will be much fewer and farther between.

We'll leave this section with a quote from Stephen Pinker [6]: 'There is not the slightest reason to believe in a coming singularity...sheer processing power is not a pixie dust that magically solves all your problems.'

Zero-Point Energy

This bizarre creation has cropped up in the chapters on both the science of the very large and very small so here we're going to look at it in a little more detail in the hope of finally putting it to death. If you are already of sick of this and reckon you've got the point then feel free to jump to the next chapter.

Though the Heretic would say that zero-point energy is an absurd beast and deserves to die, it is nonetheless rather a pity as some have proposed using it as a limitless source of non-fossil fuel burning energy.

The quantisation effect of particles, such as electrons in atoms, dictates that they can only take certain fixed values and this is the basis of quantum mechanics. Zeros make unwelcome appearances in the maths that underpin the theory. One such is zero-point energy which, despite its name, produces its own bizarre force.

The zero makes its unlikely appearance in Heisenberg's Uncertainty Principle which says that the more accurately you know a particle's position, the less accurately you know its velocity and vice-versa. This principle comes about because every observation we make perturbs the thing we are measuring. Even if you are making measurements of something by looking at the light coming off it, the tiny pressure of the photons of light on the object will make it move. In the macroscopic world this hardly matters but at the particle level it does.

Zero-point energy arises if we constrain a tiny volume of space so that what few particles are in it cannot leave. The tinier the constraining box we use the more accurately we know the position of the particles. However, because of the Uncertainty Principle, that means we know less and less about their velocity and, in turn, their momentum and their energy.

With this greater uncertainty comes greater fluctuations of the energy, and as the volume of the constraining box goes to zero, the fluctuations in the energy supposedly becomes infinite. Of course, this is only in theory as we can't make a zero-sized box. It's a ridiculous idea, but because some physicists confuse the perfect world of maths with the imperfect universe, this nonsensical concept supposedly has a vast bearing on the nature of reality.

In any case this doesn't make sense because vacuum is empty, right? But if you believe in those nasty zeros and infinities, as almost all physicists and mathematicians do, then this means that it definitely isn't empty. Suddenly the vacuum must be seething with vast and indeed infinite energies.

And it gets worse because according to Einstein, mass and energy are supposedly equivalent. So, along with fluctuating infinite energies comes fluctuating infinite masses. In other words, the empty space is supposedly boiling with infinite particles popping in and out of existence.

To try and give some patina of credence to this concept, physicists have christened these phantasms 'virtual' particles. And this fluctuating infinite energy has been bestowed the title of 'zero-point energy'.

Physicists in the 1940s ignored these ramifications when they were applying the equations and acted as though zero-point energies didn't exist. That's sort of like pretending an elderly aunt hasn't just farted at the family Christmas Party. Unfortunately, some experimental data was ascribed as supporting the actual existence of this effect. The idea behind the experiment had been proposed by Casimir and Polder in 1948 but would be very difficult to measure as it involved tiny forces between two metal plates. Nevertheless, despite the difficulties this tiny Casimir effect was believed to have detected by Lamoreaux in 1995. Results were within 15% of expected values.

Sounds sort of believable except when you look at it more closely. Casimir and Polder originally intended to look at something else called the van der Waals force which would also cause forces between two plates (it is also supposedly the force which allows Geckos to climb sheer surfaces). A chance conversation with Niels Bohr suggested there might be another zero -point related force at work as well.

In fact, the van der Waals force is actually a kind of catch-all expression to sum up various other forces of attraction in this situation. These will include three forces you've probably never heard of: the Debye force (force between permanent and induced dipoles), the London Dispersion force (between induced dipoles) and the Keesom force (between permanent dipoles). Over and above these forces contained within the catch all van der Waals 'force' there are other known mechanisms and forces of attraction between, for example, hydrogen bonds, between covalent bonds, plus other electrostatic interactions.

So even without our 'virtual particle' effects we have at least six other forces contributing to the possible tiny attraction between two very close parallel metal plates. Perhaps there are also other forces we haven't yet recognised, forces far more prosaic than infinite particles bubbling in and out of existence everywhere and all of the time.

It might be more believable if the forces involved were large but in fact they are very, very small and closely related to the separation of the plates which have to be separated by a matter of only a few atomic diameters.

But, no matter how ridiculous it all sounded, it fitted the theory and by accepting it as the truth, physicists never had to be troubled again by those infinities and zeros cropping up in the maths of Quantum Mechanics. That solved an awful lot of problems.

However, the price they had to pay was to believe that every tiny bit of space had infinite particles being continuously conjured out of existence and then almost immediately being whisked away.

Deep down this whole idea continued to trouble some physicists. Indeed, this may be why so many jumped onto the bandwagon of String Theory because it did away with the need for zero-point energies. Of course, String Theory came with its own terrible cost. Adherents had to take the massive leap of faith to believe in many more completely undetectable spatial dimensions.

The Heretic, of course, would say that infinities don't exist in the real world and so there are no infinite zero-point energies. Measurements made of the Casimir Effect would be more likely due to experimental artefact or our incomplete understanding of the many other forces that come into play when the atoms of the two plates get so close together. It also means there are no Black Holes and no hyperspatial wormholes. The Heretic thinks that the Technological Singularity and the Rapture of the Nerds are bunk.

The Heretic's view is unpopular in our present age and not just amongst scientists. We are bombarded by Hollywood blockbuster movies and TV series and books where travel between the stars is almost a given. If there are no sneaky wormholes, or other esoteric concepts arising from physics theories dependent on the reality of the concept of infinity, then travel between the stars is highly unlikely ever to be possible.

In an age where the horizons and the prospects for man can seem limitless, where reputable bodies give money to institutions like the Singularity University, then the Heretic's ideas can appear bleak.

But the Sceptic isn't troubled by that sort of pessimism. He may feel uncomfortable with zero-point energy and may even accept that the actuality of Black Holes is simply our present best guess based on limited data. However, he believes that new discoveries will be made that will provide technologies to take us to the stars in much faster times than presently appear feasible.

Meanwhile the Believer watches Star Wars and dreams of things to come.

5

The Great Physics War and other Skirmishes

Science, after all, is only an expression for our ignorance, of our ignorance.
Samuel Butler [1]

Physics is about as unified as the Middle East. Like regions of the world with many different tribes and cultures, the field is full of sub-specialities that do not really communicate and indeed seem happy to lead quite separate existences.

This co-existence is generally peaceful and is maintained by each sub-speciality pretending that, more or less, the others don't exist.

But as with the Middle East, where one tribe borders another, there is always the potential for territorial disputes that can lead to war.

Perhaps I'm resorting to hyperbole by using the word 'war' when it comes to physics. Bearing in mind that physicists are ironically not renowned for their physicality, perhaps a better word would be 'scuffle'.

The Great Physics Scuffle Between General Relativity and Quantum Mechanics: Truth the Only Casualty

As we've seen, Einstein's General Theory of Relativity is about the very big. In essence, this says that mass warps space. Again, we go back to the analogy of a rubber sheet representing space, with a big heavy bowling ball in the centre. This weighs the sheet down, forming a depression (or 'gravity well') in the centre. Place a tennis ball near one side of the sheet and it will roll down the slope created by the bowling ball until it gets to the centre. The bowling ball (mass) has warped the sheet (space).

All well and good, but a troubling consequence of Einstein's equations means that when an extremely massive star collapses it would create a Black Hole. Because of these underlying equations the Black Hole, as we saw in the last chapter, takes up no space at all yet continues to have colossal mass. The nice smooth rubber sheet (in other words a nice smooth continuous mathematical function) suddenly becomes a singularity which is the worst kind of discontinuity. Even Einstein wasn't happy about this despite it all arising from his own equations. Suddenly the troubling aspects of zero and infinity sullied the idealised equations.

Meanwhile Quantum Mechanics tries to deal with the very small. Unlike relativity and gravity which generally deal in smooth (continuous) mathematical functions, quantum mechanics is all about discontinuities. Particles leap from one quantum state to the next without any smooth transitions. The maths here also throws up a big problem. According to quantum mechanics, when you get close to zero distance from an electron its charge and mass become infinite and so another nasty singularity blossoms.

So, does this mean the charge of the electron is really infinite? Not when we measure it, it isn't. Some serious sleight-of-hand is going to be required here! Interestingly, the right value for our measured charge of the electron can be obtained if we measure its charge at a certain distance from the particle. So that's just what physicists do. There is no theoretical underpinning for this choice of distance, it just fits the data.

However, you cut it, this is nothing more than a fudge. It is another example of 'renormalisation' [2]. In essence, particle and quantum physicists carefully ignore this nasty infinity because otherwise their equations are nonsense.

So, both the physics of the large and small suffer from embarrassing and inconvenient infinities, but that's about the only thing they have in common. What's even more disturbing are the profound differences between the maths of the very large and the very small. The difference between smooth and discontinuous is like a huge abyss separating the physics of the two sub-specialities.

If there are underlying principles to the universe, why do we need two entirely different types of physics and indeed maths to describe them?

And the same kind of discrepancies can be found in everyday scale physics as well. For example, thermodynamics and mechanics also don't join up into a nice unified whole (see [3] for a more detailed description of this disconnect).

Rather than going back to basics and questioning whether our maths is perhaps very limited in reflecting reality, many physicists instead took an even bolder step into the land of myth in trying to unify the maths of the big and small. They contrived to get rid of the uncomfortable zeros and infinities by inventing a new dimension, the existence of which is supported by no evidence at all.

Physicists did this by believing that everything was made of vibrating Strings. Mass and also charge no longer went to infinity and there was no longer any nasty singularity lurking in the long grass.

According to String Theory all particles in the universe are the same, it's just differences in how the Strings vibrate that separate them. These Strings are about 10-33 cm in length. That's a billionth, trillionth, trillionth of a centimetre.

But does it finally provide us with a unifying Theory of Everything; the Holy Grail of Physics for a hundred years? Though strings are only one dimensional, String Theory needs another nine dimensions (or ten depending on who you listen to), so ten (or eleven) new spatial dimensions in total to sort of make it work.

And, as we have seen in the Chapter 3, none of this can be experimentally verified as the sizes and energies we would need to measure are too small and too large respectively. Studies of the very small have up to now needed the high energies of the sort the Large Hadron Collider can produce. However, strings are so much smaller than the sub-atomic particles studied in accelerators (they're the size of a neutron compared to the size of our solar system) that the energies required would take an accelerator millions of trillions of kilometres in circumference. That's quite a bit larger than the 22 kilometres length of the Large Hadron Collider which cost many billions of dollars to construct.

So, there's no way of proving whether Strings exist or not.

And it's even more complicated than that as there are a number of different types of String Theory. Just like String Theory was used to try to reconcile the physics of the very small and the very large, then the different String Theories needed to be unified and reconciled under the aegis of the M-brane theory, where branes are multidimensional membranes. And that needed yet another dimension.

But if you've already conjured another ten (or eleven) into existence then what harm is there in adding yet another?

Again, this all completely unprovable, so it's a bit like religion. Religion, however, can play the faith card that science can't (or at least shouldn't, String theorists please note) because if something was provable you wouldn't need faith

This leaves us with several unpalatable choices. Do you prefer a theory which posits at least ten entirely new spatial dimensions for which we can never find proof but which at least goes some way to bringing the very large and the very small under one mathematical roof? Or, alternatively, that entirely different mathematics are required to describe the very large and very small which rather flies in the face of the idea that there is an underlying unified theory to explain everything. Or perhaps you are beginning to wonder if our maths and theories don't represent reality anywhere near as closely as we think.

Up to now in this book we've looked at some of the absurdities underpinning aspects of the studies of the very small and the very large, in other words cosmology and quantum mechanics. However, similar problems are inherent within all aspects of physics.

At best the 'laws' of specialised physics areas, such as electromagnetism or thermodynamics or mechanics, apply to idealised objects in situations where other factors have supposedly been eliminated. They also generally do not translate from one speciality to the next. Thus, the core problem with the notion of physics being a unified science is that it simply isn't. As time goes by the situation gets worse as the science splits into smaller and smaller sub-specialities, each with their own sets of 'fundamental' laws. The laws of general relativity don't transfer down to the quantum level; laws of electrical resistance don't transfer down to very low temperatures. And of course, the fundamentals of quantum mechanics don't transfer up to the macro world of human beings, unless you really believe a tree hasn't fallen in the forest until you actually see it lying there.

If laws were really fundamental then why is there the need for this cascading compartmentalisation into a multitude of sub-specialities? So-called fundamental physics laws are not truly fundamental because they do not always apply. Instead they apply only in specific cases restricted by size, temperature, velocity, pressure and so on.

Fundamental laws therefore do not describe reality just as, as we saw, mathematics also cannot describe it. Though both can take us a long way when it comes to manipulating nature, the fact they are based on unrealistic assumptions will always set a limit on just how far we can actually go. These laws and the mathematics that underpins them are simply analogies. They're useful but sooner or later, like all analogies, they will always break down

This book has so far dealt with maths and physics and will later will deal with aspects of biology, medicine, engineering and the social sciences, but there is little about the subject of chemistry. Chemists may take offence for what I am about to say next and may argue that I, as a physicist, am taking rather a parochial approach to their subject. However, that subject is based around atoms and molecules which operate, supposedly, using the laws of physics. In a sense chemistry is kind of a sub-speciality of physics but it's worth making a couple of specific points about it.

The rough rules of chemistry are known from observations, in the sense that if you add substance A to substance B then you will get, under certain circumstances, a substance C. But the reactions which cause this depend essentially on the behaviour of electrons which, if our quantum mechanical theories of physics are correct, we should be able to accurately model.

Except that we can't. All our quantum models regarding the electron have been made from experiments where particles are considered in isolation. Unfortunately, this does not represent what is actually happening in nature. In isolation, the maths of these particles is relatively simple. However, when it comes to the down and dirty reality of the multiple electrons in the shells of a variety of atoms interacting within each molecule then it all gets way too complicated.

Quantum mechanics therefore can't model how atoms and molecules will react.

Recently, in order to try to make the maths of the complex interactions in biological molecules work, a new hybrid theory has been constructed. This involves part Newtonian mechanics and part quantum mechanics. The Newtonian approach basically models the molecule itself but the actual parts that do the chemical reacting are modelled using quantum mechanics. Martin Karplus, Michael Levitt and Arieh Warshel won the Nobel Prize for Chemistry in 2013 for this work [4].

This mix and match approach to physical principles rather flies in the face of their supposedly fundamental nature. The fact is that neither type of physics fits the whole thing.

Yet chemistry, based as it is on actual experiments with real chemicals, seems to have got along fine without the need for a close tie-in with quantum mechanics.

We have to be careful here about what laws and principles we are actually talking about as there is the need for some distinction. Theoretical laws purport to explain reality, but phenomenological laws simply describe what we can observe but don't attempt to describe any mechanism that underpins it.

This book is arguing that the phenomenological laws of subjects like Chemistry are really the nearest we can get to the truth in that they describe regularities in how the universe usually operates. An example would be the observation that increasing a voltage over a simple circuit increases the current. That phenomenological law doesn't say how or why this happens, it just says that it does.

Nancy Cartwright, the philosopher, and not the actress who is the voice of Bart Simpson, in her demurely entitled book: How the Laws of Physics Lie [3] says that our theoretical laws are simply a simulacrum. The Oxford English Dictionary definition of this term is: 'something having merely the form or appearance of a certain thing, without possessing its substance or proper qualities'. To her our 'fundamental [explanatory] laws do not govern objects in reality; they govern only objects in models'.

This is the last chapter that confines itself to the subject of physics and some of you may be relieved to hear this. However, before we leave it, it may be useful to look at a couple of examples from history which show how current theories have developed. This is to try to give a different slant on the commonly held view that the march of science is unstoppable; that it provides a single logical viewpoint which cuts through to the reality of the universe. It's never as straightforward as that.

In a conventional history of nations and their rulers, the winners generally write the history books and the losers, if not written out entirely, are usually forgotten. The same is true in science. The process to accepted wisdom is actually far muddier than the usual historical accounts suggest.

For example, an important point to understand is that the theories that fall by the wayside often do so not because the predictions they make are found to be incorrect, but rather because their assumptions do not match the current scientific orthodoxy. Another major confounding factor is that theories based on fundamental misapprehensions can still make very good predictions.

A prime example of this is James Clerk Maxwell's nineteenth century theory of electromagnetism in which he used mathematics to unite light, electricity and magnetism.

During Maxwell's time there was a great deal of experimental data from Michael Faraday and others concerning electrical and magnetic phenomena but there was no general theory capable of explaining them. Maxwell's unifying theory met with some opposition as physics up until then had been based on relatively simple nuts and bolts mechanical models. Scientists at the time did not accept that electric and magnetic fields were even real things.

Maxwell's great insight came from his observation that electromagnetic waves and light had the same velocity. As an aside, it is interesting to point out that Maxwell's classic equations in fact have two solutions, one where light moves forward in time and another where it moves backwards. This latter solution has always been ignored because...well...it just doesn't fit.

In essence, Maxwell believed that electromagnetic waves were vibrations of the aether. In this way he was able to explain, and so make predictions about, refraction, reflection, interference and diffraction of light. Revealing as this did the frequency basis of electromagnetic waves, this also suggested the possible existence of waves outside the frequency range perceivable at the time, like radio waves and x-rays. This was long before mankind had the means to verify their existence using experiment. The fact that one day we did detect these surely means the theory was, and is, correct; that it has 'passed the test of time'?

Maybe not.

Brilliant stuff though it is, and it's still taught in textbooks, what is not often understood is that Maxwell's work was based on two fundamental assumptions which are nowadays rejected by current physics orthodoxy. The first assumption is to do with the aether which Maxwell's equations required the waves to travel through. The second assumption was that there was an absolute frame of spatial reference, in other words some fixed point in space from which all motion could be measured.

As we have seen, strong doubts were cast on the existence of the aether by the experiments of Michelson and Morley. These in turn were the inspiration for Einstein's Special Theory of Relativity which also did away entirely with the possibility of there being an absolute frame of reference.

Now Maxwell's equations have since been modified by others to square this circle and fit it in with the current orthodoxy but the fact is that the original theory, though fundamentally flawed from a modern perspective, produced some amazingly good predictions. How did it manage to do that?

Theories in science are subject to selection based on a sort of Darwinian survival of the fittest. In fact, for every 'successful' (in other words currently acceptable) theory there are a multitude of others which seemed reasonable at the time but now, when remembered at all, are often thought of as silly or naive. They have become forgotten because they produced predictions which could not be replicated or because, even though they did explain phenomena well, they just didn't fit any more with the general scientific orthodoxy of the day.

Whatever the currently accepted theory of anything is, it doesn't mean it is right. Rather it simply means it has been cobbled together to best fit the known facts at this point in time and also to fit in with the current general orthodoxy. When conflicting new experimental data comes in then some theories, like Maxwell's, can be carefully manipulated to reflect the new facts or views, but otherwise they are junked entirely.

So, let's be careful before we start sanctifying theories that have 'passed the test of time'. Generally, they haven't done that at all.

In fact, if you develop enough theories to explain a set of facts then some, by chance, will make predictions about future experiments which subsequently prove to be correct. However, as more and more data come in, then finally the old theory's luck runs out and a new one, or a modification of the old one, has to be introduced to explain away the new findings.

Wars are difficult to gain perspective on while they are still happening. To try and gain some insight into some of the misapprehensions that may dog our present state of knowledge, we're going to look back at a physics skirmish that many believe is long since over and the winner declared.

This example also shows how the understanding of one particular physics phenomenon has developed. It illustrates the range of sometimes bizarre theories developed to try to explain it based on the scientific orthodoxy and belief systems of the time. It will also illustrate how much chance, rather than the predictions from theory, play a role in discovery and how supposedly inviolable laws may be cast aside in the desperate search for an explanation for observations. In short it shows just how messy and haphazard the whole process is. This messiness is generally expunged from the physics textbooks to leave an apparently simpler and more coherent advance along the path to knowledge where everything seems so obvious in hindsight.

The example is the historical development of our understanding of the phenomenon of radioactivity. This has been investigated for over a hundred years and so gives us a long enough view to pick our way through the maze of conflicting theories that have been put forward to explain it. It shows how wrong great scientists can be at any given time. Hopefully this will give us some perspective on how wrong scientists may be today.

Many of the following historical details come from the excellent book 'Radioactivity' by Marjorie C. Malley [5].

Some of these old theories about radioactivity will appear crazy to our supposedly enlightened viewpoint which is itself laden with its own truly bizarre concepts like Dark Energy and multi-universes and all sorts of new spatial dimensions. How we regard scientists investigating radioactivity a hundred years ago may well be how scientists of the future will regard the bizarre theories of today.

In 1895 the discovery of x-rays and, in 1896, the discovery of other new rays from uranium (what we now call gamma rays) was utterly unexpected. The science of the day gave no predictive inkling of their existence.

The prevalent idea then was that of the aether, an invisible weightless fluid that permeated the whole universe. Indeed, some believed that matter itself was made of particles of this aether. The concept of the aether (Greek for 'pure, fresh air') had been around for over 2000 years. Maxwell, as we have seen, believed that the light waves described in his equations travelled through it, as did Christiaan Huygens in his wave theory of light. Dmitri Mendeleev, the originator of the Periodic Table of the elements, even in his 1903 revised version, included aether within it. In fact, he considered aether made of two other gases, a heavier one called coronium and the lighter one which was (then) conventionally identified as aether.

At that time many other scientists also believed in 'energetics' in which everything, even matter, was made of energy. Matter was merely a ripple in this sea of energy.

This sea of invisible energy, and the existence of the aether, led many at the time to believe that this ghost world was where the souls of the dead fled to upon death. Mediums, people who could supposedly span the worlds of matter and the hidden worlds of energy and aether, were thought to be sensitive to the perturbations in this world and were sought out to enable communications with the dead. Many figures of the scientific establishment became converts to this 'spiritualism' including Alfred Russell Wallace the evolutionary biologist, Charles Richet the Nobel prize winning physiologist, and William Crookes the chemist and physicist. Even Nobel prize winning Pierre Curie, one of the fathers of radioactivity experimentation, showed great interest in these ideas.

Hard as it may seem today from our loftily enlightened view but spiritualism at that time was not regarded as a fringe idea and actually flowed from these physics-based concepts.

Modern physicists may scoff at the idea of aether and coronium and the spirit world. However, they will then quite happily go back to their cats that are both dead and alive at the same time, or their infinite number of universes.

Fair's fair though: the scientists of a hundred years ago would be equally dismissive of the bizarre views of contemporary physicists.

Anyway, back in the late 1800s the world was suddenly full of these newly discovered rays, including also infra-red and ultraviolet as well as a number of mysterious others. These included cathode rays which came off a negatively charged electrode and which would later go on to help form the images on early TV sets. The chemist Sir William Crookes went so far as to postulate that these cathode rays were a new fourth state of matter. In other words, they were neither liquid nor solid nor gas.

Rontgen had discovered x-rays by chance in 1895 when, while trying to eliminate all light from his laboratory for a cathode ray experiment, he found that fluorescent material near a device that produced cathode rays was itself glowing. This was despite the fact that the device was covered by black cardboard and was too far for the cathode rays (which are electrons) to reach it through the air. This suggested some other new type of ray was penetrating the solid of the cardboard and passing through the air almost unaffected.

Soon the penetrative ability of these rays was being exploited. They were found to pass through flesh more easily than bone and so could be used to form pictures that were essentially density maps of the human body. The rays were call x-rays because their cause was unknown at the time.

Gamma rays were also discovered by accident. Antoine-Henri Becquerel was investigating how x-rays were tied in with fluorescence and was looking for new materials that fluoresced. If they could be made to do so when exposed to light then he hoped they may in turn emit x-rays.

This actually isn't possible but the misconception led him to use uranium salts as they had previously been found by Becquerel's father to brightly phosphoresce. This led him to believe it might be even better as a conversion device for changing light to x-rays.

To Becquerel's surprise, the uranium somehow blackened photographic films even when there was no light at all. In other words, the uranium was producing highly energetic radiation all by itself.

However, he was still convinced this was all just a hangover from the material's earlier exposure to light. In order to overcome this supposed delayed effect, he kept samples of uranium in the dark for over a year. Even then they still darkened the plate. Still not convinced, it took many more experiments before he began to consider that the rays came from uranium as some form of 'special case'. In other words, he was still hanging on to his theories regarding light producing x-rays when shone onto certain materials.

The scientific community at the time generally believed that the gamma rays were simply a type of x-ray. As x-rays were much more easily produced then research into uranium rays fell by the wayside. The phenomenon that would lead to nuclear energy and the atom bomb became, for a while, a research dead end in physics.

Modern physics has it that both x-rays and gamma rays are electromagnetic waves that are produced by different mechanisms. X-rays are produced from electron shell energy transitions, gamma rays from transitions in the nucleus of the atom itself.

However, Marie Curie didn't believe gamma rays were a dead end and was soon investigating their ionising effects on gas. This meant they gave electrically neutral atoms an electrical charge.

She discovered that as well as uranium, another element called thorium produced these by a process she christened radio-activity in 1898. She found that the relative amounts of uranium in different ores like pitchblende (meaning 'black appearing and containing metal') and chalcolite, did not match the degree of radioactivity she detected.

Nowadays we believe this is due to the relative amounts of the different uranium isotopes present in the ore. Isotopes are versions of an element with different numbers of neutrons in the nucleus. Curie did not even consider this as she was hidebound by prevailing scientific dogma. For more than a hundred years it had been strongly believed that elements were immutable and each had a fixed atomic weight. This dogma was partly a result of the appalled reaction to the previous hundreds of years of alchemy. Alchemy held that by creating something called the Philosopher's Stone it would be possible to convert base metals into silver and gold, and also to create an elixir of immortality.

Interestingly, one adherent of alchemy was Sir Isaac Newton who some would argue was the greatest physicist of all time.

However, by the time of the Curies, alchemy had fallen very much into disfavour so respectable scientists really didn't want to go there. Rather than having truck with the idea of atoms changing their nature, Marie and her husband Pierre Curie instead took the view that the radiation came from other elemental impurities. In trying to further isolate these impurities they discovered two new elements: polonium and radium. This was an entirely fortuitous discovery because these new elements were in fact not the cause of this mismatch of radioactivity.

There is at any given time rarely a single theory to explain any results and indeed there were other explanations for what the Curies had found. Some scientists suggested that radium was actually just a uranium impurity and it was that that was producing the radiation. Another theory suggested that just as iron can be magnetised by contact with a magnet, a substance might be made radioactive by previous contact with uranium.

To prove they really had found a new element, the Curies tried to show that it had a different atomic weight to other elements. This led to the extraction of the tiny amounts of radium impurities (one tenth of a gram) from large amounts of pitchblende (about 100 kilograms). The ensuing radiation exposure from the extraction process led eventually to Marie Curie's death and probably would have done the same for Pierre had he not fallen under the wheels of a horse-drawn cart.

Ernest Rutherford believed these rays were a form of light and he investigated them in those terms but could find no evidence of reflection and refraction. This, according to the theory of the day, ruled them out as a form of light. He also found that they contained other types of ray which generally fell into two categories: one that could be stopped by aluminium foil and one that could penetrate several sheets of it. He called the former alpha rays, the latter beta.

In terms of the highly penetrating component of gamma rays the consensus was they were a form of x-rays and not light (in modern theory all three are considered types of electromagnetic radiation).

Marie Curie thought that all of space may be full of a high energy radiation that, on striking uranium and being absorbed by it, produced the gamma rays. Thinking this mysterious radiation came from the Sun, the Curies repeated their measurements at different times of the day. At midnight, they believed the mass of the Earth would attenuate more of the radiation from the Sun than when it was directly overhead.

That theory had to be discarded when they found no significant differences in the results whatever the time of day. Instead, they began to suspect that the law of conservation of energy, that most hallowed of physical laws which underpinned the whole universe and was thought inviolable, might be incorrect.

Others, like Sir William Crookes, proposed that uranium somehow gleaned energy from the motion of air molecules against it. Some like Adolf Heydweiller, Robert Geigel and Sir Arthur Schuster thought it somehow gained the energy from gravity.

Meanwhile respected researchers Julius Elster and Hans Geital were proposing that radioactivity might affect the weather. The tie in with weather also led them to believe that heat and pressure would affect the radiation produced by uranium, but again this was not supported by later experiment.

They tested the Curies' theory that radiation from some other external source than the Sun caused the rays by taking their sample down a deep mine in the Harz Mountains, but again no change was found.

Another theory was that, as x-rays and gamma rays were the same then maybe bombarding uranium with x-rays would produce gamma rays. It didn't.

In fact, numerous experiments over many years by many researchers could find no means of increasing or decreasing the rate of production of gamma rays.

If the precipitating cause for this radiation wasn't from energy being supplied externally, then it all must happen in the fabric of the uranium itself.

In 1898 Rutherford proposed that radiation came from a re-arrangement of the structure of the uranium atom. Another theory was that the atom, rather than being a solid object, was composed of parts that moved. At certain times the moving parts would form an unstable whole at which point the atom would disintegrate and the energy would power the radiation emitted.

It must be emphasised, however, that the concept that this breakdown changed the nature of the atom itself was never considered. Nobody wanted to be tainted with the stigma of being labelled an alchemist. The idea that one radioactive element would change to another element in the process was still years away.

Few of this multiplicity of theories that were eventually discarded make it into current text books of physics, so the transition from ignorance to our present theories of radioactivity seems much smoother and quicker than they really were. One is instead left with the impression that a few simple theories, tested by careful experiment, steadily took us to where we are now.

For every new finding there are multiple theories. Only the ones that keep fitting new data as it is produced are retained, but that does not mean the surviving theory is correct. For example, nowadays radiation is considered in purely probabilistic terms. In other words, if you took a single radioactive atom you could never say exactly when it would decay. You can only say what probability it has of decaying in a certain time interval. But is that really true?

Our Heretic would take this intrusion of chance as a sign of the failure of our present theories of radioactivity to understand the actual process. If the theory was realistic and correct then why shouldn't it be able to predict exactly when an atom would decay?

Even today the issue of radioactive decay is much more of a mystery than is commonly appreciated.

This book has suggested that one of the few, indeed perhaps the only underlying truth in the universe is that no two things in it are identical. It seems to be true for everything we see, from little things like grains of sand all the way up to big things like galaxies. However, physicists seem to regard all atoms of the same element as being entirely identical and, similarly, all sub-atomic particles such as protons to be identical to all other protons. Why it should be the case for objects on the atomic scale and yet nowhere else in the universe is difficult to imagine.

But the fact is that atoms of an element created at the same time do not all decay at simultaneously. Rather than consider that there is something non-identical about them to explain this, scientists instead resort to the concept of probability. Elements are considered as having a probability of decaying in a certain time for no reason which is explained. Not that scientists haven't looked for an explanation. Possible variables that might affect radioactive decay include heat, pressure, and gravity. None have been found to affect the rate at which radiation is emitted from an atomic nucleus. Indeed, no other physical effect has been found to affect spontaneous decay.

Few scientists consider the recourse to probability as representing a huge gap in our theory. After laying down so many supposedly immutable, unchangeable scientific laws, it's quite a failure to have to resort to admitting that a physical process is random. Resorting to probability is like putting your hands over your eyes and pretending there isn't a problem at all.

There have been theories to explain radioactive decay that didn't resort to statistics. As we've already seen, Elster and Geitel over a hundred years ago thought that the reason a given nucleus produced radiation was that the atom was made of smaller parts in motion. At a certain point they found themselves in an unstable configuration, though at the time Elster and Geital did not go so far as to suggest the atom actually broke down.

Sir Oliver Lodge and J. J. Thomson later suggested that it was when one of the moving parts fell below a certain velocity that the system broke down, the way a spinning top falls over when its spin rate drops below a certain level. They went further than Elster and Geitel and suggested the unstable atom may then even explode.

Rutherford thought that the electrons slowly drained energy from the atom. This was testable because it meant older atoms were more likely to decay. In the end this was not verified by experiment.

Soddy and later Debierne (1912) used an analogy with the theory of gases, where sometimes individual molecules may interact to give additional energy to another; enough sometimes for it to escape from a liquid as an evaporated gas. Later Friedrich Lindemann proposed that rather than particles in the nucleus moving, it was their spins that interacted to cause decay. Marie Curie likened an atom to a box with a little hole in it by which a moving nuclear particle might sometimes escape.

At least these early scientists had not given up on there being an underlying physical mechanism that would explain why a given atom disintegrated at a given time. Somehow, over the last hundred years, probability has become an explanation of the decay process itself, rather than being a sign of failure. Indeed, nowadays quantum mechanics has to deal very largely with chance and probability to supposedly explain all sorts of sub-atomic processes.

One last point before we leave the nuclear world. It's a common misconception that nuclear power and nuclear bombs arose from predictions made from theory. In fact, it all came from the chance discovery by Pierre Curie who noticed that his radium samples were always several degrees above room temperature. If it was true that the additional heat came from the radioactive disintegrations he measured, then they produced an awful lot of energy; enough, in fact, to raise the temperature of its equivalent weight in water from zero to boiling point in only an hour. It was this observation that led Rutherford and Soddy to suggest that the energy stores in the atom could be the source of energy in the Sun.

So, the present orthodoxy is that all the measurements we make at the quantum mechanical scale are supposedly smeared out by the uncouth intervention of 'chance'. But at least there is nothing wrong with the way we make measurements in the human scale everyday world, surely?

Sadly, that is not the case. For centuries we have been measuring all sorts of things but generally only recording the results we expected and ignoring the rest. That's the subject of the next chapter where we're going to start looking at some of the fundamental limitations that dog all science.

6

Chaos and Prediction: Reality Comes Knocking but is Ignored

Science may be described as the art of systematic oversimplification.
Sir Karl Popper, philosopher

In the 1960s the Western world's confidence in science and technology was booming and it seemed inevitable that the universe would one day be tamed by mankind. Rather than being confined by ropes and chains, the dangerous universe would be brought to heal by the equations deduced by our brilliant scientists. The apparent vagaries of reality would be cut through by the methods of science. From the carefully elucidated mechanistic processes that underpinned the universe, we would be able to accurately chart its course and the course of everything within it.

That said, prediction would not necessarily be easy because of the vast amounts of calculation required but we were then at the start of the computer revolution. This new and increasingly powerful and increasingly cheap technology would give us the muscle to map out the future of everything.

But cracks soon began to appear and undermine this dizzyingly exciting vision. They appeared first in some of the 'softer' sciences like economics. With the recent availability of powerful number crunching computers, economic and employment forecasts were produced for companies and nations and for world trade. Predictions were being made to supposed accuracies of up to three decimal places.

Unfortunately, the actual final outcomes would bear little or no relation to these predictions however many decimal places they went to.

It got worse though, because even the simplest of physics equations, taught and unchallenged for hundreds of years, also seemed to fall apart when they were used to predict how any system would behave. Even simple experiments with oscillators, such as pendulums, which should be eminently predictable, just didn't go the way they were supposed to [1].

Some people even began to suspect that something had gone wrong with the physics; that perhaps the laws had changed over time. In order to check whether this wild idea had any validity, the raw data from historical experiments in mechanics were reviewed. It turned out that the same thing had always been happening. Experiment had never been producing the results that theory had predicted. As far as the anomalies in the behaviour of, for example, pendulums were concerned this was evident as far back as Galileo and his initial work on them.

Why hadn't this been noted at the time? It turned out that these results had simply been dismissed as anomalies. After all, the theory of mechanics was so elegant and simple and self-consistent that it had to be right. So, these anomalous results had been put down to glitches or experimental error or equipment malfunctions or poor calibrations. Some were dismissed as outliers. These were data points so far off the trend that they just had to be considered incorrect.

In the case of pendulums scientists had dismissed the anomalous results because the mechanical theory governing the pendulums was so conceptually obvious that, if the experiment didn't progress as predicted, then something had to be wrong with the experimental setup.

And thus, for hundreds of years, an important truth was ignored.

So even these simple predictions had proved incorrect but, of course, prediction is more typically used to calculate the behaviour of much more complex systems. For example: the orbits of planets, the spread of epidemics, how biological and economic and meteorological systems change with time.

Unfortunately, both simple and complex systems have something in common and it was this something that served to undermine both.

All predictions based on models ranging from the simple (pendulum) to the highly complex (the weather) have to have a starting point. A computer (though in earlier times it was just a man with a pencil and a piece of paper) will then typically apply the model to calculate how the system evolves from these starting conditions.

For example, if you are looking at the new stocking of a lake for the purpose of breeding trout you may begin with just a pair of fish. From the rate at which such trout have been measured to breed, and assuming enough food is available, then it is apparently a simple matter to calculate how the population will grow with time. Provided there are no external factors such as the introduction of disease or a new predator, the model will show a smooth population increase until it reaches a plateau level because at that point the food available is just large enough to support the final stable population. There may be a slight overshoot to begin with and the overall population might go up and down a bit for a while, but the level will eventually converge to the steady plateau value.

That's the theory, at least.

The above example uses the simplest starting condition, namely one male and one female fish, but it should be possible to do your predictions at any stage, for example if you start the simulation when you have a thousand fish. Both simulations should end up with the plateau population value where the availability of food sets a limit on final population size.

Again, it all seems so obvious.

Unfortunately, the truth isn't anything like that. Generally speaking, the population never reaches a semblance of a stable plateau value, but instead varies wildly.

Essentially the same kind of modelling process is used for meteorological predictions where the starting point is the weather as it appears right now. Such simulations are much more complex than the simple fish stocking example because it is subject to a wide range of pressure and temperature conditions in the atmosphere. It is the multitude of interactions of these regions of higher and lower temperature and higher and lower pressure that will govern how the system will evolve. This is why computers have become so necessary in meteorology. But despite nowadays using some of the most powerful computers on Earth, weather forecasts are not highly accurate. Indeed, it might be argued that the biggest boon that computers have given to meteorologists is an awareness of just how inaccurate their forecasts are. Back in the 1960s the BBC and the Meteorological Office would issue long term weather forecasts for the coming month. Better computational analysis of the accuracy of these forecasts revealed their predictive accuracy as being no better than chance. Nowadays most meteorologists confine their predictions to little more than three days ahead and the accuracy of these, as we will see, is subject to question.

So, what's going wrong with our predictions, both complex and simple?

In 1961 Edward Lorenz was running a simple weather prediction programme. Wanting to repeat some measurements, but without doing the whole lengthy simulation again, he used results from the first run except he started the simulation midway through rather than at the beginning. The predictions from that point on should be the same for both but, to his surprise, he found they were very different.

Looking more closely, Lorentz found that the problem lay in a computer printout of a single variable used in the mathematical modelling. The computer was using a value of 0.506127, but when this was printed out it was only to three decimal places. In other words, it appeared as 0.506 on the printout and that is the number Lorentz had used to re-run the programme. The difference between the two numbers is trivial, on the face of it. The difference is less than a fiftieth of one percent. According to the thinking of the time it shouldn't have made any significant difference to the predictions.

But what Lorenz stumbled upon was that even the simplest models were highly sensitive to the initial starting conditions. Even tiny differences in these could make the simplest predictive models behave very strangely.

Nowadays, after this field of study was dubbed 'Chaos Theory' and became something of a vogue in the late 1980s [1], the extreme dependence on initial conditions was given the popular name of the Butterfly Effect. The idea behind this was that a butterfly flapping its wings in Brazil could supposedly cause a tornado in Texas.

Systems which change with time, whether made up of only a few or a multiplicity of variables, like pressure and temperature at various locations in meteorology, can be disrupted by even the tiniest changes in one or more of these.

Simplicity had descended into chaos and once people became aware of this they started to see chaos everywhere, in biology and engineering and physics. Equations for predictive behaviour quoted in textbooks that had been used for generations turned out to be wrong. Systems which were considered to inevitably come to a steady state given time never actually did.

And of course, it wasn't that the models and equations had suddenly stopped working in 1961. They had never worked yet nobody had seemed to notice.

When people reviewed the older scientific literature it became clear that this had always been the case. Even as far back as the 1880s Henri Poincare had found an example of how little our cherished physics equations represented reality. Poincare was concerned with what is called the three-body problem. Let's take as an example the Sun. A planet orbits round the Sun, and there is a moon which orbits round the planet. Newton's laws of gravitation should allow us to calculate how the bodies interact gravitationally and thus predict how they would orbit around each other.

What Poincare showed is that there is no simple, regular, repeating solution to the equations. They never reach a stable value. Make a prediction of where the bodies will be in a thousand years based on where they are today and you will get one solution. Make the thousand-year prediction based on the position of the bodies in a year's time and it will be entirely different. In other words, the trajectories are unstable and unpredictable and so chaotic.

This has profound implications for the fate of the Solar System. It means that because of this underlying complexity, and despite increasingly accurate observations made over several hundred years, the whole system is not stable in the long term. It is possible that in time the planets' orbits will become increasingly eccentric and that one day they may even leave the Solar System entirely [2].

That physics cannot cope with accurately predicting the motions of three interacting bodies is a truly fundamental limitation. The universe is full of systems which involve countless such interactions, whether they involve planets or weather systems or economies.

Of course, science almost throughout its history has tried to shut out the complexity of reality. Everything in the universe is perturbed by more than one other thing but science struggles mightily to set up experiments which exclude as many of these as possible so they can study the 'pure' effects. For example, if you want to study the force of gravity here on Earth you may try to measure its force by observing how quickly objects fall. If possible, you would try to carry out the experiments in as near vacuum as possible so you would greatly reduce the effects of air resistance, wind, the effect of temperature on the density of the air and so on.

This scientific process of trying to reduce the confounding effects of other variables is called reductionism and we will discuss it in the next chapter. At this point it is enough to say only that we can never eliminate reality's multitude of confounding effects.

So, in just fifty years, we've gone from a belief that the universe was something that would ultimately be amenable to hard, accurate prediction, to a new understanding that reality is actually unpredictable except in very, very carefully controlled circumstances and, even then, not completely.

This new, more humble awareness marked a huge retreat by science. It clearly signalled its inherent limitations. For a while it was proposed that Chaos Theory was itself the way forward and that perhaps within it was a deeper order that could be manipulated to enable prediction. In the 1990's if you wanted your paper published in almost any field of science, or if you wanted your grant application to be funded, then you were wise to put the word 'Chaos' somewhere in the title.

But over the last twenty years those high hopes have dribbled away and all we are left with is an awareness of the limitations of our science and its predictive capabilities.

The effects of chaos (a word used in the specific sense of the unpredictable effects on systems that are highly sensitive to initial conditions) did not spring into existence with the discoveries of Poincare or Lorenz but have been lurking in the data in almost all fields of science since the dawn of prediction. It's just that, somehow, they were ignored. One wonders what else science may be presently ignoring. After all, even today, faith in the absolute truth of theory is still widespread. And if results don't match the orthodox theory then there will always be a tendency to believe the results must be wrong.

But let's look now at some of the more common modern-day attempts at prediction and try to assess how useful they are.

Meteorology

People have been trying to predict the weather at least since the Babylonians in 650 BC. Weather has always been an obsession of mankind as its effects can be so catastrophic or bountiful. In the developed world it might only mean a spoilt Bank Holiday or an extended commute; in less developed countries it can often be the difference between life and death.

Before computers came along for mankind to put its misplaced faith in, all meteorology was based on observing and remembering weather patterns and then remembering what happened next. For example, observing clouds, winds, changes in temperature, and then later recognising similar patterns.

The earliest form of prediction, and one to which earlier man would be exquisitely attuned, was the pattern of the seasons and the weather it brought. The further towards the poles you live, the more incontestable the effects of the seasons are on weather.

The bedrock of weather prediction is the collection of data. How can you tell what the weather is going to do next where you live if you don't know what is happening now one hundred or a thousand miles away? The invention of the telegraph and radio allowed the necessary transmission of temperature and pressure measurements. It allowed forecasters to get an indication of what may be coming their way.

The first steps were also taken to harden up the quality of the data. Identification of cloud types became based on standardised 'cloud atlases' and the levels of wave disturbance caused by the wind became defined by the Beaufort scale.

However, it was not until 1922 that Lewis Fry Richardson first made the suggestion that there may be a quantitative basis for weather prediction, though even then it was recognised just how demanding the computation process would be. Indeed, it was not until the 1950's that sufficient computing power became available to at least begin the attempt.

Data in the form of variables like pressure, temperature, precipitation and wind velocity are now transmitted from automatic weather stations, satellites and from airline pilots and crews on ships. The more densely sampled the information, and their rate of change, then the better the estimates of the initial conditions for any predictive model. These are then fed into the 'primitive equations', as they are called, which attempt to project these ahead in time.

All these equations, however, are non-linear and based on rates of change. These hardly ever produce unique solutions. Textbooks, incidentally, are full of differential equations which have unique solutions. This can lead the unwary to believe differentials are usually and uniquely soluble. In fact, they are most definitely not.

This non-uniqueness of solution means that slightly different models produce sometimes quite markedly different results. This is why it is finally up to a meteorologist, based on his own experience, to make the final choice between competing computer predictions.

Trying to determine the accuracy of weather forecasts is something of a dark art in itself and perhaps not surprisingly as there are a multitude of weather conditions at a multitude of places to measure at any one time. You can prove anything, depending on what and where you draw your sample. However, some interesting studies have been done [3] which looked at the accuracy of predicting rain (defined as at least a tenth of an inch of precipitation) for five Kansas City meteorologists over a seven-day period. Accuracies ranged from an average of 85 percent for the first day but fell to 73% by the seventh day.

On the face of it that doesn't sound too bad, but one should bear in mind that if the weathermen had always predicted it would never rain on any day, they would have scored an accuracy of 86.3%. Three of the forecasters beat this, but only just, and only on day one. In all subsequent days they fell below this value.

Meteorologists' own estimates of when accurate prediction becomes essentially impossible ranges from three days to ten days ahead depending on which one you listen to.

Computing power has increased more than a million-fold since 1960. How much has the accuracy of weather forecasting improved in that time? The answer is a few percent, if that. Yet still the meteorologists clamour for more powerful computers and greater data collection.

Biology and Ecology

'Natural balance' was a hallowed concept in studies of species populations throughout most of the twentieth century. Population ecologists such as Robert MacArthur, Frederic Clements and Henry Cowles held it as an article of faith that natural (as opposed to artificially man-managed) populations would find a suitable equilibrium (or 'climax' as Clements called it), where food and water and other resources were just sufficient to sustain a stable population.

It was a fine idea which in the end turned out to be purely theoretical rather than realistic. Species are affected by a multitude of factors. There is the issue of predation, often by a number of different predators. The populations of these different sorts of predators will themselves be affected by the availability of food provided by the different populations they prey on. These in turn may be affected by the availability of food to the species they prey on, whether they are animal or vegetable. Vegetation, in particular, is affected by weather which is itself unstable. Diseases which affect the species also come and go in at least cyclic and often chaotic manners.

The result is a multi-factorial, non-linear differential equation where small changes in just one variable can have unpredictable effects on all these interacting species. Modelling populations, it turns out, is fiendishly complicated. If you want to see examples of nearly sixty sources of error in population modelling see [4].

And, of course, when the new awareness of this underlying complexity dawned on ecologists they found it everywhere in the data. 'Natural balance' had been such a compelling concept that previous experiments that had showed no such natural balance had all been dismissed as failures.

In all aspects of biology, when we look for simplicity all we find instead is complexity, no matter how convincing and comforting our theories may be. As we will see later, DNA-as-blueprint has been the recent most simplistic biological theory that appears to be falling apart under closer examination. Like 'natural balance' we think we will find simplicity, only to be overwhelmed by complexity.

Economics and earthquakes

Billions of pounds have been spent on modelling economies and also the performance of stocks and shares. Some of the cleverest minds and most powerful computers are devoted to this modelling process.

Earthquakes can be utterly devastating as the early years of the twenty first century have shown all too clearly. The 2004 Indian Ocean earthquake killed nearly a quarter of a million people, the 2011 Tohoku earthquake and tsunami killed around 18,000. Again, vast resources have been devoted to trying to predict earthquakes.

How successful have been these attempts been at predicting the future? Life's too short to go into any detail about either of these but I'll simply give you the following quote from the statistician Nate Silver:

'In November 2007, economists in the Survey of Professional Forecasters--examining some 45,000 economic-data series--foresaw less than a 1-in-500 chance of an economic meltdown as severe as the one that would begin one month later. Attempts to predict earthquakes have continued to envisage disasters that never happened and failed to prepare us for those, like the 2011 disaster in Japan, that did.'[5]

Summary

The universe is innately complicated, indeed this book makes the case that complexity and uniqueness are perhaps its only truth. The awareness of the effects of 'chaos' has given the lie to the belief held for hundreds of years that systems and events and even the universe itself followed a deterministic path. 'Chaos' is an example of the intrusion of this complexity to undermine the deterministic universe envisaged by scientists like Newton and Laplace and many others.

It gives the lie to the belief held for hundreds of years that very small influences could be neglected; that predictions of future behaviour might not be exact but they would be good to a high level of accuracy.

And no amount of increasingly elaborate maths or faster computers will help us.

7

The Human Brain and DNA: The Limits of Reductionism

The problem of science for Nietzsche is ultimately a problem with the prioritisation of a generalised intellectual approach to existence - in its separate parts and as a whole- that holds no object to be immune to its conceptually grasping gaze, understanding, and manipulation, an approach to reality that looks at all things from the outside, as an observer does, rather than from the inside, as one who lives in reality does.

David Taffel [1]

Reductionism is the method by which hard science progresses. Faced with the bewildering complexity of reality, it tries to break down even the most complex processes or things like the human brain, into bite sized chunks each of which can then be experimentally investigated and its workings analysed. Each little bit becomes like a piece of a jigsaw puzzle. Finally, at the end of this reductionist process, when all the little bits have been individually understood, then they can supposedly be added back together until we can finally understand how the whole complex system works.

Except in reality that last bit almost never happens. We'll look at the human brain and human genome as examples, both subject to intense reductionist investigations, where this final step in the process is still as distant as ever. This suggests that perhaps the reductionist approach may itself be a major limitation of the scientific method.

That complex objects are simply the sum of their constituent parts is an idea that has been around since Descartes in the seventeenth century. This was contrary to the much older Aristotelian view that the whole was more than the sum of its parts. Nowadays this latter view has had something of a resurgence and is exemplified by the words holistic and emergent. Holistic refers to the idea of treating the whole thing, for example an animal, as an integrated functioning organism rather than cutting it up and analysing its constituent parts. Emergent expresses the idea that new properties emerge out of the complexity of the interaction of all the little reduced bits which could not have been predicted by just studying these parts individually.

To many, holism is associated with aspects of fringe medicine and all the wackiness that ensues, from reflexology to aroma therapy and beyond.

Emergent properties on the other hand are beginning to find a greater respectability in science primarily as a result of the growing awareness of the limitations of the reductionist approach in explaining how complex systems manage to work. For example, individual neurons may have been investigated to within an inch of their lives but this has yielded no insights as to how the whole brain, made up of billions of these, can appreciate beauty or find Laurel and Hardy movies funny. Somehow these higher functions 'emerge' from the cooperative activity of the myriad single neurons. Where scientists reach for the concept of emergence is a sign that the reductionist approach has failed.

Emergence is seen when a collection of objects self-organises, usually because of inherent feedback mechanisms. In other words, it has no central co-ordination. Examples include situations where crowds of individuals start to riot, or where the behaviours of investors can cause boom or bust in financial markets, or in the development of hurricanes and typhoons. Life, according to this view, can be seen as the emergent behaviour of groups of molecules.

Prediction is only possible where emergent properties are suppressed; for example, where the feedback between individual units is negligible, or where there is enough variation in individual behaviour to wash away the beginning of an emergent signal.

This lack of predictability of group phenomena gives rise to uncertainties in trying to study it, as is so often the case with the social sciences. Historians, sociologists, economists and those involved in the study of politics can only offer interpretations of events rather than hard analyses to predict future behaviour.

As ever, when faced with complexity, people try to see order and this has led some to suppose that emergent phenomena themselves have universal properties. Indeed, the study of these has given birth to a new branch of science called Complexity. Some have even lumped this in with Chaos Theory. Horgan, the Believer's champion, in his book 'The End of Science' called it Chaoplexity [2].

However, there is supposedly a difference between complexity and chaos. Chaotic systems vary so erratically they can seem random, whereas complex systems can have pockets of self-organising order.

Either way, despite much publicity in the 1990s, neither Chaos Theory nor Complexity has yielded anything of substance in the way of new scientific methods or discoveries.

Incidentally, in this book you are reading, complexity is used just to mean that everything is different, rather than being used to describe complex systems which produce emergent behaviour.

But despite the recent need to resort to the idea of emergence, the reductionist approach has always held sway for hundreds of years now. For example, when early doctors and biologists wanted to know how the human body worked, it was broken down for study into its constituent parts and functions. The muscles for movement, the blood for carrying nutrients to the organs, the heart for moving the blood round the body, the skeleton for stability and protection of the softer organs, the gastrointestinal (GI) tract to digest and eliminate food, kidneys and livers to eliminate toxins from the blood etc. Organs could be extracted, dissected, viewed under microscopes and chemically analysed. Knowledge of these individual organs or systems was built up over many years in this way. Understanding of disease was usually achieved by correlating the diseases that a patient suffered from in life with post mortem investigation of their tissues and organs.

This knowledge of organ function was greatly supplemented by invasive experiments on living animals, the animals being used as models of the human.

Inevitably, as more and more knowledge was gained about individual organs, then medicine split into specialities. Gastroenterologists study the GI tract, cardiologists the heart and so on. And it's not just the anatomy that causes specialisation, but also the physiology of the body. Physiology is the way the different organs and systems actually work, and this too has generated different specialities such as endocrinology (study of the endocrine glands and the hormones they produce).

Similar consequences of the reductionist approach can also be found in physics where specialism causes an increased chopping up of the subject into smaller sub-units like mechanics, thermodynamics, particle physics, cosmology, optics and so on. All have their own cadre of specialists. As we saw previously, each speciality such as optics or thermodynamics produces new theories and laws which don't seem to have much common ground with the others.

The impetus for taking the reductionist approach is quite understandable. After all, you're hardly likely to discover how the universe works in just one experiment. But the trouble is that this approach, rather than demonstrating that there exists a common foundation on which everything else is based, is that we see instead fragmentation. Indeed, fragmentalism is the less polite term for reductionism.

The more we experiment using reductionist methods, the more specialities we create and the greater the disconnect between them.

Let's now talk about two of the biggest challenges to our scientific understanding, whether you are into reductionism, holism, Chaos, Complexity, Chaoplexity or whatever.

The Human Brain

Where to begin! The human brain is central to everything because it is the filter through which we try to understand the universe and all that is in it. To many it is the seat of the soul, though other times and other cultures have considered the heart the repository of this. One of the main functions of the brain is supposedly to enable us to perceive the universe. Indeed, if the purveyors of the Copenhagen Interpretation of quantum mechanics are to be believed, using this mighty organ to observe a quantum phenomenon collapses the whole universe into a single state. After all, the much-misused Schrodinger's cat is both alive and dead until a human observes it.

Reductionism is everywhere when it comes to the study of the brain and has been applied intensively for over a century with hundreds of thousands of neuroscientists and others devoting their entire careers to the problem. Yet what has all this effort shown? How does the brain actually do all the fancy things it does?

It's worth looking at the history of where our knowledge of the brain comes from.

Neuropsychology is a speciality which tries to marry the structure and function of specific areas of the brain with behaviour and other psychological processes. The speciality of neuropsychology more latterly dates from the French doctor Paul Broca in the latter half of the nineteenth century, though as early as 3500 BC the Egyptian physician Imhotep had started the ball rolling by looking at the effects of trauma to the brain on the victim's subsequent ability to function. His ideas didn't find much traction with his fellow Egyptians who instead would go to great lengths to preserve the heart after death because they believed that this was the seat of intelligence. They also even preserved the viscera after death so the person could keep using them in the afterlife but regarded brain tissue as no more than head stuffing and so tossed it out as being useless.

The Ancient Greek philosophers were more divided on this issue. Plato thought the brain was the seat of the soul, whereas Aristotle thought it was primarily used for temperature control.

Not much happened for the next two thousand years but then the brain became the focus of concerted study in the twentieth century. Some of the first studies to determine which areas of the brain were responsible for what functions took place in the First World War. In this, many fit young men were hit in the head by high velocity metal in the form of bullets or shrapnel. Such serious head wounds were, and indeed still are, very often fatal. However, in the Great War there were so many men and so much metal flying around that there were quite a few survivors. Neuropsychologists would apply a battery of tests to them to see which aspects of brain function had been affected. They would test for such things as language and physical coordination, memory and sight, emotional stability and reading capability.

Loss of function could then be correlated with the site of the wound but the really high-quality data came from those experimental subjects who later died and whose brains could then be carefully studied. The areas of the brain that had been damaged could then be correlated with any neuropsychological deficits that had been detected with the tests on the person carried out while they were still alive.

At the time this was the best data that could be obtained but obviously there are some serious problems with it. Firstly, the assumption was usually made that before the soldier had suffered his injury that he had been 'normal', whatever that is. It was assumed that he could read and write, that his hearing and sight were reasonably good, that he was moderately coordinated, was of average intelligence, and so on. However, in the First World War recruiters weren't too choosy and so these assumptions may often not have been correct. In other words: any deficit may have been the result of nature and not of high velocity metal.

A second problem is that bullets and shrapnel don't generally enter the head, destroy one specific area of the brain, then exit politely. In fact, bullets, and the shock wave that often comes with them as they move through tissue, will affect all the different areas of the brain. This made the task of correlating affected brain regions with functional deficits all the more problematic.

Nevertheless, areas of the brain having some involvement in vision, language and motor control could be reasonably well identified. Further elaboration became possible by producing lesions in specific regions of the brains in animals and studying the effects, though obviously this was of limited value in determining the effects on higher cognition. For example, asking a rat who the last three Presidents of the United States were will not get you very far.

However, some of the most significant advances were made by investigating human subjects with small, naturally occurring lesions in very specific sites of the brain. The advent of high resolution non-invasive imaging techniques, such as Computerised Tomography (CT) and Magnetic Resonance Imaging (MRI) in the latter part of the twentieth century allowed accurate identification of the position of these lesions without having to wait for the subject to die and be autopsied. Indeed, nowadays subjects with small single lesions in specific parts of the brain are in great demand and can make a living travelling the world to be tested at specialist brain research institutes.

A more recent elaboration of the MRI technique is called functional MRI (fMRI) and this has allowed study of the blood flow changes in parts of the brain stimulated by some event. For example, if you show a subject a flashing checkerboard, then the blood flow over areas of the brain responsible for visual processing will become stimulated and blood flow to that area will increase. This allows us to visualise areas of the brain that are responding to a certain stimulus. However, there is a significant limitation because these blood flow changes take seconds to be detectable. This is much too slow as many cognitive functions happen on a much shorter time scale. For example, if you recognise a face in a crowd you usually do it in a fraction of a second. That's too quick for the sustained blood flow responses needed to show up on fMRI and other blood flow related imaging techniques such as Positron Emission tomography (PET).

Other ways of investigating brain function non-invasively include electro-encephalography (EEG) and magneto-encephalography (MEG). These have much less certainty about showing the position of the part of the brain being activated by a task but do show the very rapid changes associated with the electrical firing of the brain cells.

The invention of the microscope and the use of specific staining chemicals have also allowed the brain's micro-structure to be revealed. Silver chromate was first used by Camilo Golgi in the 1890s to show individual brain cells (neurons). These cells, found in the grey matter of the brain, came to be regarded as the key individual elements responsible for brain function. They are thought to do this by transmitting electrical signals and also chemicals across junctions between them, called synapses. In the human brain there is something like 90-100 billion neurons, with each neuron typically making synaptic connections with 7000 others. Somehow, the theory goes, higher cognition, mind and soul are engendered by the scale and complexity of these interconnections.

Interestingly some worms have as little as 300 neurons, fruit flies 100,000, ants about a quarter of a million, honey bees a million.

In fact, the whole issue of insects and their brains is an interesting one as it casts some doubt on the fundamentals of our understanding of the human brain. Carl Linnaeus, the father of the taxonomic classification of creatures into different groups, in the eighteenth century constructed his insect grouping on the basis that they had no brain at all. He based this on the observation that insects such as horseflies can live after decapitation. The Comte de Buffon, another naturalist of distinction at the time, went so far as to say the '...horsefly, will live, run, nay, even copulate, after being deprived of its head'.

There is even some recent evidence that parts of a cockroach's body other than the brain may be associated with memory [3]. Cockroaches can survive several weeks without having a brain at all.

Insects also give us pause when it comes to our understanding at even the most basic neuronal level. Insects, despite in some cases having only a millionth of the neurones we ourselves possess can still do quite a lot. Some have hundreds of more eyes than we do but somehow their limited number of neurones can integrate them all into an overall view of the world. They also have antenna to feel, which we generally don't, and their tiny brains need to take on board the stimuli coming from the numerous hairs on their body which allow them to sense the wind. They also sometimes have wings and, if you've ever tried to swat a spirited fly, you'll know what exquisite control they have of these.

They also know how to hunt and trap prey, how to eat it, how to mate, to recognise certain smells, how to navigate to and from their nests.

And some insects can build truly impressive things. Even though termites have relatively few neurons, they produce vast and labyrinthine citadels up to ten metres in height. They build roads, bridges and tunnels across the forest floor and all with two lanes for the two directions of travel. Below their mounds they excavate down to sixty metres to find water.

Charles Darwin [4] was very impressed by insects and by ants in particular:

"It is certain that there may be extraordinary activity with an extremely small absolute mass of nervous matter; thus the wonderfully diversified instincts, mental powers, and affections of ants are notorious, yet their cerebral ganglia are not so large as the quarter of a small pin's head. Under this point of view, the brain of an ant is one of the most marvellous atoms of matter in the world, perhaps more so than the brain of man."

Not bad for a creature with only about a quarter of a million neurons.

In order to explain how creatures with such small numbers of neurones can accomplish so much, scientists resort again to the concept of 'emergence' in respect to behaviour of the creatures as a group. For example, somehow, out of lots of individual insects, complex behaviours such as swarming appear. Analogies are made between each ant and each neuron of the human brain. Working together the ants supposedly form a kind of meta-brain, like the individual neurons make up the human brain.

The truth is that these are rather hand waving arguments for something we really don't understand.

Let's look at another big problem with our understanding of the brain.

How do we remember things? The truth is that nobody knows despite years of intensive investigation. Scientists have, as ever, broken down memory into different types (short term, long term, working, sensory etc) and studied the effects of various external challenges on these. They have also identified specific areas of the brain which seem to be involved in the laying down and retrieval of memories. For example, tiny lesions in areas of the brain like the hippocampus and amygdala can have catastrophic effects on short term memory. In affected patients, recent memories are lost within minutes.

Indeed, I remember one such poor gentleman who was a patient at the neuroscience institute where I worked. Every day his wife was continually and repeatedly having to explain what was happening to him or what he was trying to do because he would forget within minutes. Nevertheless, he was still able to do his work as a postman because his longstanding postal route was lodged in his unaffected long-term memory.

So, there are clearly areas of brain tissue that are important in the laying down and retrieval of memories, but where in the brain are memories themselves actually recorded? Studies of the effects on memory from brain injury raise more questions than answers. While some concussions can cause memory loss, particularly for the things that happened just before the injury, the loss of long term memory is not related to the amount of brain tissue destroyed. Indeed, sometimes neurosurgeons will quite deliberately cut out quite large volumes of the brain, for example to treat types of epilepsy not responsive to drug therapy. Neurosurgeons excising tumours can take out even larger volumes of the brain. It's been found that even removing a third of the brain tissue may be associated with no appreciable memory loss. If a third of the brain has been removed then why aren't a third of memories lost?

An even more striking example can be found in the case of patients who have lived for many years with undiagnosed hydrocephalus. This is a condition in which more fluid is produced in the brain than is absorbed. Slowly, over the years, the brain tissue is displaced by this cerebrospinal fluid. Such patients may come to the attention of neurologists because they are troubled, for example, by minor but persistent symptoms like a 'muzzy' headache. When imaging is performed it can reveal that as much as 90% of the patient's brain tissue has been displaced by, essentially, water. Even so, many of these patients continue to do their jobs, raise their families and be members of society but with only 10% of their brain tissue.

Where these patients manage to store their memories is certainly a big question but it pales into insignificance beside another: why are these patients still even human?

Such cases are very unusual but there are other types of pathology where brain tissue is greatly reduced that appear on a weekly basis at neurological hospitals. Low grade brain tumours called gliomas can grow so slowly that over the years the brain can, up to a point, compensate for this growth and so not produce symptoms. When finally presenting at hospital, perhaps with the sudden onset of fits or even just the intractable muzzy headache, it's found that most of a brain hemisphere may have been replaced by tumour.

Yet these patients are still the human beings they were and their memory is usually unimpaired. How can this be?

Conventional neurological explanations include the concepts of brain plasticity and redundancy. In the case of plasticity, if an area of the brain is compromised by a lesion of one kind or another, the brain can sometimes rewire itself over time so that other areas assume those functions. Motor recovery after stroke would be one example where this can happen to a degree. In the case of huge lesions or cerebrospinal fluid spaces, where large volumes of brain tissue are lost, the idea is that the brain can still function because much of it is redundant in the first place.

But the idea of having bigger brains than we need (perhaps by as much as 90%) for the purposes of redundancy doesn't make sense from an evolutionary point of view. More redundancy means bigger brains that in turn mean bigger skulls. Before the Caesarean sections that modern medicine provides, big headed babies killed expectant mothers and the babies themselves in their droves. Also, their big heads mean that human babies take a long time to be able to move to get out of danger. Compare this to a newly birthed foal that can walk and even run within minutes. Human babies will therefore be much more vulnerable to predators. Is all that extra mortality of otherwise healthy babies just to give better chances of survival to rare cases like brain tumour (about one in six thousand per year) or long term undiagnosed hydrocephalus (a tiny number of cases per year) or severe head injury (most of whom would die anyway without modern surgical intervention to decompress the swelling brain)? If we really needed less brain tissue then smaller brains would have a survival advantage from an evolutionary point of view to both the babies and their mothers.

But let's get back to memory. Where memories are stored is still a mystery. Scientists, as ever when things get difficult, have to reach out for analogies. If they can't find a specific area where memory is stored then that leads them to the conclusion that memory must be stored on a 'distributed' basis. The analogy used is often that of the hologram. Even small areas of a holographic plate hold information about the whole image so that not all information is lost even if the plate is broken into fragments.

The actual mechanism for this supposed holographic storage of memory is not known however; it's just a guess.

An alternative explanation is based on the old mind-brain duality way of looking at this problem. Hundreds of years ago Leibnitz and Descartes were driven to adopt this concept of dualism; that the body and the mind were two completely separate entities. Dualists might argue that the brain can be thought of as the interface between the mind and the body, and that memories may be stored somewhere else entirely in some kind of metaphysical iCloud.

But, of course, dualism really doesn't fit within the prevailing we're all just molecules materialistic school of thought of the present day, as exemplified by the books of evolutionary biologist Richard Dawkins [5].

Neither the holographic nor the dualist view is backed up by hard evidence. Where memories are actually stored is as much of a mystery now as it has always been.

The effort put into the study of the brain is gargantuan and has been sustained for many years. Over 12,000 papers are published every year just on the subject of fMRI alone. More generally, scientific papers investigating the central nervous system appear at a rate of over 60,000 per year with the total number of neuroscience related papers in reputable peer reviewed journals coming in at over 1.5 million. These figures are continuing to grow exponentially. The Society of Neuroscience alone has 42,000 current members.

And what has been the outcome of this vast experimental effort? Has the reductionist approach of studying separately all the different areas and functions and microscopic architecture of the brain helped us explain us. Has it explained how 1.5 kilograms of tissue can produce art and literature, how it can love and grieve, how it can appreciate humour, how it can feel tenderness, how it can remember or even just think?

The truth is that all this vast effort hasn't even come close to explaining how something that looks like a solidified mass of porridge can produce thoughts and emotions.

Faced with these levels of ignorance, a blunderbuss approach is being taken to delve into the secret workings of the brain. The BRAIN Initiative (Brain Research through Advancing Innovative Neurotechnologies) was launched by the US government in 2013 and is believed to be costing $3 billion over ten years. The idea seems to be to map all 80 billion or so human neurones. Scientists will somehow do this by using technologies that haven't even been invented yet, hence the hefty price tag. Meanwhile, the European Union's Human Brain Project includes one billion Euros to build a model of the brain using computers. About 7000 man-years of effort has been budgeted for this. The show-runner for this project, Henry Markram, used a TED conference speech to indicate the ambition of creating a self-aware artificial intelligence [6]. This new creature/machine will supposedly allow the brain to be fully understood up to, and perhaps even including, consciousness itself. Markram somehow hopes to integrate the vast amounts of data generated by all the different fragmented fields of neuroscience. It is hoped that some simple unifying principles will emerge from this data to allow the group to cut through the apparent overwhelming complexity.

To the Heretic, resorting to such blunderbuss approaches is suggestive of desperation, of a reductionist approach that has singularly failed to help us comprehend how the mind works. The brain has at least 80 thousand million neurones, with perhaps a hundred million million connections, and at each of these connections electrical activity is modulated by dozens of different neurotransmitters acting through over three hundred different types of ion channel. Faced with this overwhelming complexity, is it really surprising that despite all this time and high-powered effort, with so many scientists spending whole careers on their reductionist approaches of studying only one cell or molecule or channel or whatever, we are really no nearer understanding how the brain is capable of higher cognition. We have no idea how it makes our minds.

There are many sceptics of the Human Brain project with over 130 leaders of scientific groups threatening to boycott it as they consider it 'radically premature' and 'doomed to failure' [7]. We'll leave this section with the words of one such sceptic, Christof Koch, Chief Scientific Officer at the Allen Institute for Brain Sciences in Seattle: *'The round worm has exactly 302 neurons, and we still have no frigging idea how the animal works'* [6].

Now let's look at another mystery where science appears mired in cascading levels of complexity.

DNA and the human genome

The unravelling of the DNA (deoxynucleic acid) molecule has been the Holy Grail of Biology for the last seventy years. DNA is considered the basis of all life and its long strings of molecules called genes are believed to be the recipe for human beings and for all other life on Earth, or at least what science recognises as being life. Genes are regarded as the molecular units of heredity and dictate the structure and function of the organism.

What better target for reductionist science than the investigation of long chains of coded instructions?

Children inherit some features, called traits, such as blood type or eye colour from their parents but the mechanism for this transmission had been a mystery. DNA was first identified in the pus from used bandages in 1869, but it was nearly a hundred years before experiment appeared to confirm that it was the 'giant hereditary molecule' first proposed by Nikolai Kolstov in 1927. In 1953 James Watson and Francis Crick proposed, from an x-ray diffraction image, that the molecule was in the form of a double helix.

DNA molecules are huge, the longest is 220 million units in length and each one would measure 8mm if it wasn't scrunched up tightly within the cell. Current genetic theory has it that the long DNA chains consist of strings of four molecules called nucleotides (adenine, guanine, cytosine, thymine) which form the basis of the genetic code. Computers work on a binary system of zeros and ones, but the DNA is essentially based on these four units. In other words, it's like a computer whose basic units would be 0, 1, 2 and 3.

Additionally, sequences of three of these nucleotides combine to form one of twenty types of amino acid. A number of amino acids combine to make a gene which in turn produces a protein via other molecules called RNA and ribosomes. There are perhaps 20,000 different types of protein in the human body and perhaps a million in total in all living things. One protein can be made up of as many as 26,000 amino acids. Proteins are involved in biochemical reactions in the body and transport other molecules into cells, as well as building and repairing the cells themselves. Different proteins help build different types of cell such as in the brain, skin, liver, heart and kidneys.

When the egg from the mother and sperm from the father combine the genes become mixed. There can be over 400 genes, each contributing a genetic trait, in each of the 23 different chromosomes in humans which make up the DNA. Although the child will have the same number of genes as each of the parents, some will come from the father, some from the mother. That is why children are never identical to either one of their parents. However, things are a little more complicated than one gene being responsible for one genetic trait. For example, in the well-studied fruit fly (Drosophilia) at least fourteen separate genes affect eye colour.

These numbers are already getting rather large but nonetheless the theory seems, at first glance, relatively straightforward. Nevertheless, there are a number of mysteries. Here come a few examples.

The first one is that it appears that only about 1.5% of the total genetic sequence in humans actually code for proteins. This compares to 98% in some bacteria. Of the remaining 98.5% of the genetic code in humans, about half is apparently non-coding repetitive sequences, sort of like the chorus in a ballad. Why is it there? Some of these apparently redundant bits can be explained away as being molecular fossils, genes which have been disabled during the evolutionary development. For example, since our far ancestors crawled from the sea we no longer need the genetic code to make gills.

Much of this supposedly redundant or otherwise useless code was originally dismissed as 'junk DNA', however, increasingly, functions are being
found for it. For example, parts of it may somehow dictate the rate at which
other parts of the molecule produce proteins.

Nevertheless, there is considerable disagreement in the literature about
how much of this DNA actually performs no function at all.

Another mystery is the size of the genome in terms of the number of
nucleotides. This can vary over 3000-fold between animals, and 1000-fold
between plants.

And as humans are at the top of the evolutionary tree, the pinnacle of the
evolutionary process, then surely we'll have the greatest number of nucleotides?

Not even close. The Marbled Lungfish and some salamanders, for example, have about forty times more nucleotides than we do. In fact, except in the
case of 'simple' organisms like bacteria and viruses, this nucleotide size does
not seem to bear much relation to the number of genes. The organism with
the most DNA is not even an animal at all but rather the lily which has at
least ten times as much as we do.

Another problem comes in identifying the boundary between one gene
and another, or genome annotation. It's not straightforward and is often
based on statistical rather than definitive identifications. As we have seen
when it comes to physics, resorting to statistics may be a sign of simplistic
theories breaking down.

But, leaving aside the above mysteries, what we appear to have here is a
recipe for each individual human being. By analysing the sequence in his/her
DNA chains it should be possible to work out the proteins created. With such
knowledge we could potentially manipulate the genome while the organism
was still at the single cell level, deleting those strings of coding that may make
them subject to some genetically transmitted illness. Or we could change
their eyes to have a more aesthetically pleasing colour, or we could make
them smarter and stronger or more like a popular media star of the time.

In short, we could make the human race into supermen and superwom-
en. What could stop us, leaving aside wishy-washy liberal concerns that this
might not be too dissimilar to Nazi-like eugenics?

There are also worthier reasons for investigating the genome, partly to do
with the search for pure knowledge, but also to help us detect diseases hidden
in the genes before they started to manifest themselves. For example, understanding of an individual's genome may reveal an increased likelihood of developing cancer or heart disease in later life. Early intervention may prevent
the disease manifesting itself.

So, if these theories (and remember they are just theories) are correct, there's a lot to play for. An intense effort has therefore been put into mapping out various genomes. The human genome was announced as being mapped in 2003, though not quite all to the original high level of intended accuracy of 99.99% [8]. The sequence also did not come from just one human being but was rather a conglomerate of sections from different people.

Though the mapping took thousands of man years and cost the huge sum of $3 billion, the cost of the technology for mapping sequences has dropped markedly since then so that rapid sequencing of an individual's genetic code could one day become as routine as a blood test.

So, what's to stop us mapping our own genetic profiles and using them to predict the occurrence of diseases like Alzheimer's, arthritis and so on, and to start preventative treatments many years before the first signs and symptoms appear?

The answer, of course, is complexity. As ever, the more we know the more we realise we don't know. This is a common statement in science and some say this shows humility on the part of scientists in their search for the truth. The Heretic would say that we are fitting more and more complex models to something which has no explanation, or no explanation that we as humans with our linear thought processes and analogy-based thinking could understand.

One index of this burgeoning complexity that has emerged from striving to match genomic code with the diversity of each human, is the genesis and growth of the '-omics'. These are whole new scientific fields that have burst into existence. Some examples:

Epigenomics (from the work epigenetics meaning 'beyond genetics').

Proteomics

Transcriptomics

Pharmacogenomics

Toxicogenomics

All of these have been spawned, or at least had their growth as a subject accelerated, by the study of DNA. Rather than DNA being a simple formula, each new specialist field reveals cascading levels of complexity the deeper we go.

The first of these, epigenomics, arose from unexpected observations that were emblematic of a greater underlying complexity than anticipated. The traditional view of genetics is that DNA, via RNA, provides the recipe for specific proteins which help construct and repair the cell and provide it with essential biochemicals. The problem is that cells containing identical DNA can be quite different in terms of their effects on organism structure, biochemistry and development. Originally it was thought this was the product of mutations in the genetic code in some of these cells, but mapping has shown this is not at all the whole story.

And it is an important story as these divergent properties displayed by otherwise identical cells affect gene expression and regulation, cell development and vulnerability to disease. In other words, just because you have a gene implicated in cancer does not mean you will develop it though it may somehow increase the risk.

Basically, identical cells may have identical genes but only in some are the genes activated ('expressed'). Many potential sources for such activations have already been identified (diet is one example) but many other as yet undetermined factors may need come into play to try to explain what we observe.

In the face of such complexity, the field again is highly dependent on the statistics and the methods of bioinformatics. This latter is concerned with organising and analysing the masses of biological data that this work generates to tease out what else is affecting expression.

Proteomics has also received a boost (and its name) from the mapping of the genome and is the study of the structure and function of the proteins created. These are seminal to the whole genetic process. Some consider proteomics even more complicated than genomics because, whereas the genome is fixed, the proteome (the entire complement of proteins) varies from cell to cell and from time to time (see the above section on epigenomics). Not only that, but proteins can become modified after they has been created by other factors (post translational modification).

Using the presence of a protein to establish a marker for disease, such as a type of cancer, is therefore obviously confounded by this complexity. It is made even more difficult by individual variations from patient to patient as well as the fact that each patient may be expressing different proteins at different times.

And it gets even worse. Single protein recipes (from the gene sequence) can actually produce different proteins by a process called alternative splicing.

On top of that, one protein by itself may not be the end of the story. It may need to combine with others that may or may not have been produced in order to form more complicated biomolecules required to perform a specific function, such as cell maintenance.

And then there's the whole business of protein degradation which depends on many other factors, including some related to other characteristics of the individual genome.

All this makes an -omics which is beset by poor reproducibility when it comes to experiment. In one study of yeast proteins 854 were identified. When the study was repeated under supposedly identical conditions 1504 were found [9].

Proteins are also a mystery in terms of how they produce their specific three-dimensional shapes. Proteins fold up in very specific ways which expose certain chemical sites and allow them to interact with other biomolecules in the 'correct' way. If they take different shapes then they can't interact and so perform their function. A typically complex protein can fold up into literally billions upon billions of different shapes. Some, it is true, require less energy to adopt and are therefore the most preferential forms, but even so there can still be thousands of shapes with the same minimum energy levels. So far, studying the chain of amino acid 'code' in the DNA has proved ineffective in working out which shape the final protein can take. This is called the 'multiple minimum problem'.

Transcriptomics deals with RNA, including the messenger RNA, that carries the code for proteins from the DNA molecule to the site where the protein is actually constructed. There are two confounds we know of (there may yet be more) which mean that the recipe may not be followed. Firstly, there is the matter of transcription attenuation where metabolic conditions can stall transcription, but the most significant cause of variation is environmental conditions. Small changes in the latter can have much amplified changes in the actual amount of specific proteins created.

So, in order to understand how the genome actually works then genomics, epigenomics, transcriptomics and proteomics would somehow have to be integrated, as well as other '-omics' that I'll spare you.

Complications on complications.

Even if by some miracle we manage to work all this out, then the issue of applying it effectively to cure disease again introduces additional complications. A subject's genetic makeup will affect how they respond to drugs. Pharmacogenomics is the study of this response. It is estimated that at least 80 percent of current prescription drugs are affected by the presence, or absence, of specific genes.

The hope of pharmacogenomics is that, by correlating expression of specific genes with drug efficacy and toxicity, and then comparing this with the individual patient's genome, we may be able to tailor the drugs to the patients, or at least not bother giving them drugs that would be ineffective.

Held out as one of the great hopes of genomics, 'personalised medicine' has not proved itself to be anywhere near the boon to mankind it was touted to be. This is a consequence of some of the inherent complexity described above.

Predictions of disease appearing in an individual using '-omics' are rarely, if ever, cut and dried, but instead are based on probabilities and statistical analyses. The situation is made worse by the fact that it may take many different activated genes to produce disease. For example, how many headlines have you seen along the lines of 'scientists find gene responsible for cancer'? That just means scientists have found some sort of statistical correlation between the disease and a single gene but that is never the whole story. 108 genes have so far been found to be associated with schizophrenia alone [10].

All of these complex processes are occurring simultaneously and this complexity of interaction can never be modelled or controlled in the laboratory. Inferences can only be made from very restricted circumstances. Affecting one process with a targeted drug could therefore have manifold unpredicted effects.

'Personalised medicine' is disappearing in a statistical haze and remains a myth.

It was thought that mapping the human genome would revolutionise medicine. It would provide the chemical programming to build proteins which in turn would allow us to build better human beings and cure the sick. Once we had mapped the human genome then we would know everything. However, instead of nailing down our understanding of what makes us, and everything else that has DNA, it has revealed only greater and greater complexity.

So, as we saw when it was faced with the overwhelming complexity of the human brain, science has again resorted to the blunderbuss approach in the hope that getting even more data will alleviate the problem and pave the way to understanding.

Take, for example, the proteomics problem which relates to perhaps as many as a million proteins we find in organisms. To deal with this huge number, a new multinational initiative is underway called the Chromosome Centric Human Proteome Project (C-HPP) [11]. The aim is to characterise the proteins produced by individual genes, leaving aside for the moment the rather significant confounds of epigenomics and transcriptomics.

The limits of reductionism become clear when we consider that the already complex proteomics and epigenomics and transcriptomics all represent interacting dynamic processes, each of which change from second to second in response to untold numbers of chemical and physical stimuli. As with all reductionist science, a single gene may be isolated and its effects on specific cells or organs studied in vitro in the lab, but how reflective will this be of its role(s) in the biochemical hurly burly of a living body?

Summary

After thousands of years of effort, our understanding of the human brain is still extremely limited. For the last fifty years, tens of thousands of scientists have been investigating the relationship between DNA and the organism and have been met with cascading and overwhelming complexity.

This puts into perspective currently fashionable ideas such as uploading our consciousness to computers and living forever or developing medicines 'personalised' to each individual based on their genome. These are still fantasies.

Colossal amounts of time and reductionist effort have revealed only deeper complexity when it comes to the brain and the genome.

This is all very discomforting for the Sceptic but he or she still hold onto the idea that the more data we get the more likely an understandable theory of how each individual turns out the way they do will one day be forthcoming. That one day the view inherent in science that we really are 'just molecules' whose genetic outcomes can be accurately predicted will be fully justified.

And that, essentially, is the Sceptic's article of faith.

But much of this is pretty discomforting to the Heretic as well. How does this fabulously arcane biological complexity sustain itself? Why don't DNA molecules produce monstrous chimeric creatures every time? Why don't animals and humans, based as they are on countless biochemical interactions that themselves vary with time, simply fall apart under the weight of their own biochemical complexity? It's almost like the universe, when it comes to life at least, has a general predisposition towards even greater and greater complexity which is somehow self-sustaining.

It's all very well the Heretic saying, for example, that two planets are attracted to each other because they 'just are', and not because of some underpinning mathematical law, but why would a universe that 'just is' produce 60,000 different metabolites and perhaps a million different kinds of protein?

Answers on a postcard, please.

8

The Limits of Observation: Donning the Blinkers

Whether you see a thing or not depends on the theory which you use. It is theory which decides what can be observed.

Albert Einstein [1]

In any act of observation there is choice. In the simplest type of observation, you have the choice of looking up or down, right or left, forward or back. As soon as you look in one direction you are not looking in the others.

In modern experiments, whether it is in the fields of physics, biology or medicine, the choice of what to observe is always critical. The more advanced the experiment and the smaller or more specific the effect to be observed, the more things we are not observing [2].

Let's take cosmology as one example. One can't observe the full electromagnetic spectrum from radio waves up through the frequencies to ultraviolet, visible light, infra-red and microwaves and highly energetic gamma rays in any single detector. The more sensitive the detector, the smaller the range of wavelengths that can be measured and the more expensive it is to make. This expense is often magnified by the need to launch these highly sensitive detectors into space to overcome the confounding effects of the Earth's atmosphere.

That means that when one is testing a theory, for example that a certain type of star will produce radiation at a certain wavelength, you use a detector sensitive to such a wavelength. The star might be emitting radiation at all sorts of other wavelengths, and so perhaps undermining your theory, but you won't know about it. You will find what you were looking for and therefore the theory, on the face of it, will be considered correct. Only when further and contradictory data comes in will you know your initial theory was wrong, or at least needed 'renormalisation' or some other fudge to make it seem to work.

In biology you may want to test for a certain protein that your theory leads you to expect to be expressed by a certain gene. You generally look for that protein and not for others.

Another example of this restriction is when we look at small structures within an organism using light microscopy. The greater the magnification we use the finer the structure we see but the less of the overall organism we can observe. The more closely we observe something, the more of the object we exclude from the act of observation.

This is true in terms of size but also in terms of time, for example if we are investigating something that changes over time like the blooming of a flower. We can generally, for reasons of expense or the amount of analysis required, observe (sample) phenomena in finely spaced intervals over small periods of time, or in much coarser intervals over long periods of time.

Not only are we inherently limiting what we observe, and so effectively closing our eyes to the complexity of the object under investigation, but we are also usually having to alter, or at least unnaturally control, the object itself. This is true in high energy particle physics where only certain particles with certain energies can be observed by specific detectors. It's usually even more true in biology. In order to view certain anatomical structures in high resolution, the confounding effects of movement of the organism often has to be eliminated by killing the creature who owns the organs. That's rather ironic considering 'biology' is defined as the study of life.

Perhaps the most obvious example of this overall problem is in electron microscopy of organisms or tissue samples. Not only is whatever we are looking at stone cold dead but the specimen has to be prepared and manipulated to an extraordinary degree before we can even attempt the observation.

These essentially denaturing operations usually start with chemical fixation. Fixation is necessary to prevent putrefaction and decay of the specimen but also to give it better strength for subsequent manipulation, for example cutting into thin sections. To do all this means inevitably introducing artefact by changing the specimen both chemically and structurally. For example, fixation of brains causes different degrees of swelling between the white and grey matter. Sliced up bits of brain in pathology museums look good but they are not the way they would have looked in the living human. On the other hand, if you hadn't used fixative, slicing the brain up would have been like slicing up partly solidified porridge.

Other forms of artefact are also introduced by the fixation process. Wrinkling of the membrane in bacteria, called a mesosome, was thought for a while to be evidence of a new functional structure of the cell but was later shown to be simply an artefact of the fixation process.

But fixation of the specimen is usually only the start of the story as far as electron microscopy is concerned. That is typically followed by rapid cryofreezing to -195o Centigrade, cutting into thin sections, embedding these into some sort of holder and staining with heavy metals which more readily stop electrons. The latter can involve spraying the object with evaporated metals like gold.

What appears on the exquisitely detailed final images that electron microscopy triumphantly produces may therefore be quite radically altered from whatever the thing looked like in real life.

As was already mentioned in an earlier chapter, in quantum physics the Heisenberg Uncertainty Principle is essentially an acknowledgement of the effect of the observer on the experiment. This says that the more we know about one parameter of a particle, the less we know about another. The more accurately we know its position the less we know about its momentum.

A more easily grasped example would be an attempt to observe where a particle actually was. Even if we were to use light to do the observation, the photons of light would impart energy to the particle. This would alter the particle's own energy and hence its velocity and momentum. This in turn would alter its position. The act of observation would inevitably alter (perturb) the very thing we were trying to look at.

On the everyday level, however, this has a miniscule effect. The energy in a beam of light does exert a pressure so that every time you shine a torch beam on something you push it a little, but the effect is tiny. Shine a torch at a bulldozer and it is unlikely to skitter away. Nevertheless, the bulldozer will move microscopically. This is the principle of solar sails where the pressure of light can give some propulsion to spacecraft. The pressure is small and only gives the spacecraft a tiny acceleration but sustained over long periods of time this can push it to high velocities.

So, in all the examples above, the act of observation is to some level confounding the very measurement we are trying to make. However, there is a deeper issue here and that is what Einstein was referring to in the quote at the beginning of this chapter. What it comes down to is that we look for what we expect to see. When the results don't comply with our expectations we interpret them still in terms of what we think we know.

Two examples illustrate this point: Rutherford's alpha particle scattering experiments and the dual slit particle/ wave duality experiment.

Let's look at the first of these. When I try to depress a key on this keyboard my finger does not pass through the key. Our bodies are apparently solid. In Greek times, it was thought that our bodies were divisible down to indivisible objects called atoms. These were little solid objects that abutted against each other like billiard balls at the start of a game. Tightly packed, they wouldn't allow other atoms to pass through them.

That made sense, but then Rutherford came along and when firing alpha particles at atoms he found that most passed through unaffected. In fact, so many passed through that he reluctantly came to the conclusion that, rather than being a solid ball, the atom was empty apart from a little nubbin of mass in the middle called the nucleus.

That was his interpretation for things he couldn't himself see or touch, being fired at objects he also couldn't see or touch. All his inferences were therefore indirect.

That the atom is almost all space was an inference based on a 'big' world experience. He interpreted the behaviour at the very small scale based on an analogy with the large scale.

The thing is, we have no reason to believe the world of the small scale bears any resemblance to our 'big' world concepts. This disconnect becomes compounded because all subsequent experiments have generally taken Rutherford's model of the atom as an acceptable representation despite the immediate paradoxes it creates. If each atom is almost all space then why don't atoms pass through each other? Why doesn't my finger pass through the key on the keyboard?

One explanation may be that electrons orbiting the atom might act as some sort of shield. This idea was itself another 'big' world analogy which didn't really work either. This was because in Rutherford's time electrons were supposed to be point particles orbiting around the atom. If this was really the case then they might well be on the other side of the atom as the incoming atom approached. One atom might then still be able to pass through another.

The way round this paradox was to consider the electron less as a particle than a wave that has a probability distribution. That allows the electron to sort of be everywhere at once within the atom. Unfortunately, this breeds another paradox in that in some experiments the electron seems to act very much as a discrete particle, as mentioned before.

Another anomaly springing from the big world analogy is that if the electron is negatively charged, and the nucleus is positively charged, then why isn't the electron attracted straight to the nucleus and smash right into it?

The way round this paradox is that, on this scale, Newton's laws of motion involving centrifugal force don't apply as the electron is a wave and has a probability of occurring anywhere, even in the nucleus. This is again a new paradox because charged objects on the macroscopic scale don't do that.

What it comes down to as far as atomic physics is concerned, is that new experiments cause new paradoxes which can usually only be solved by inventing more and more particles, more and more dimensions and more and more universes. Each time something new is postulated, experiments are constructed to detect it (except in the case of multiple dimensions and universes where scientists discard the whole scientific method and admit we can never observe them to prove they actually exist). These experiments are generally constructed to measure the energies of certain particles under highly specific conditions. If the particle is detected with the theorised properties it becomes proved even though the constraints of the experiment would have excluded many other possibilities.

The previously discussed dual slit experiment perhaps affords us a clue as to what is really going on here. If we observe a 'particle' one way it acts like a wave; if we observe it another way it indeed acts like a particle. Some physicists would have us believe they act simultaneously as both but it is the act of observation that constrains reality itself so that it behaves as one or the other.

The Heretic would say that this is absurd. The only truth of which we can ever be certain is that the universe is profoundly complex. Whatever the thing is we are observing, it is more than a particle or a wave, it is something far beyond that. We are simply taking limited viewpoints on something extremely complex and these force us to fill in the blanks of our understanding with wild and ultimately unverifiable theories.

A simple analogy might be the act of looking at your partner. Face on, they are recognisable as an object of love and perhaps beauty, but from above they're just a shock of hair, from below they are the soles of two feet. Even somehow integrating these two viewpoints won't help us understand the person; the depth of their character; the emotions they feel; the things they have done; why they put up with us.

Perhaps dual slit experiments show us only the soles of the feet and the top of the head. Depending on how we look, all we see is a hair/feet duality just as we see a wave/particle duality. We can similarly construct a mass of fantastic theory to explain this hair/feet duality but all has to be described in terms of what is possible with hair or feet and nothing else. But, of course, judging your partner based on the top of their head and the soles of their feet hardly represents their reality.

And that may be why everything in the universe, from the very large to the very small, appears so confusing.

In these last three chapters we've looked at some of the general limitations of the scientific method and the measurements we make. We've seen how so many of the simple mathematical relationships we believe govern the physical world are undermined by a startling sensitivity to the initial starting conditions. We've seen how the reductionist approach seems powerless to help us understand the higher functions of complex systems. We've seen how the act of observation is like donning a pair of blinkers, restricting our view and often only allowing us to see what we are expecting rather than what actually is.

But both the Believer and the Sceptic aren't out for the count yet. "Oh yeah," they say, "If science is built on shifting sands and subject to flaws and errors we can't even begin to understand, then how come it works? How come we have laptops and Smart phones, how come we have spacecraft that can fly to the moon and the planets? How come we can build massively beautiful and breath-taking buildings and tunnels and bridges that span continents? How come we can build cars that steer themselves and missiles that can hit a mosquito between the eyes at a range of hundreds of kilometres?"

These are all good questions and we'll try for an answer in the next chapter.

9

Engineering: The Red-Headed Step-Child of Science (and just as unfairly treated)

The myth that the knowledge incorporated in any invention must originate in science is now accepted in Western culture as an article of faith, and the science policies of nations rest on that faith.

Eugene S. Ferguson [1]

By now many readers will be asking a simple but profound question. If, as the Heretic would have us believe, scientific theories are all figments of our collective imaginations, then how come our technology is so powerful and effective? How can we construct such incredible architectural edifices as the tunnel connecting Britain to France that travels undersea for nearly 38 kilometres, or buildings like the Burj Khalifa rising out of the sand of Dubai that is 163 storeys tall and reaches a height of over 800 metres.

And what about electronics, the wonder of the late 20th century that has revolutionised all our lives: our tablets and our Smart phones, and our on-demand world of entertainment?

And what about the astonishing advances in computing power? China's Tianhe-2 supercomputer can perform a mind blowing 34 petaFLOPS; that is 1,000,000,000,000,000 floating point arithmetical operations a second.

And the rate of this development of electronics has been equally stunning. If we take the IBM 1620, marketed in 1961 as an 'inexpensive' ($64,000) scientific computer, as our performance baseline then it would take 578 billion of these to match the speed of the Tianhe-2. That's some progress in only fifty odd years. Even common household items like the Sony Playstation 4 sold for around $500 is reputed to have the computing power equivalent of nearly three billion of those old IBMs [2].

And as for computer memory, if you want to replicate the feeling of coming across something quite magical then pick up a simple USB Flash Drive. Look at it and remember that such a drive holding a very modest 256

Megabytes of memory storage has over two billion changeable physical things within it that can represent either a zero or a one. All of that on a chip which is only a couple of square centimetres.

When Arthur C. Clarke said: *Any sufficiently advanced technology is indistinguishable from magic* he was definitely onto something. [3]

And then there's the matter of cell phones. The present generation of Smart phones can tell you where you are on the Earth, can let you search for information on the World Wide Web, itself a heavy-duty bit of magic that contains a large part of the knowledge of all mankind. Smart phones can also allow you to run a plethora of applications that can do anything from displaying a star chart by simply aiming at a section of sky, to alerting you to the proximity of potentially available sexual partners.

Where do all these wonders come from if not from scientific theory? How can it be as wrong-headed as the mad old Heretic asserts?

In fact, the machines produced by our technology rarely depend on scientific theory. For thousands of years engineers have learnt from observation and experience and this has allowed them to produce novel inventions or solutions to the problems that faced them. They had hands-on experience of the materials they worked with and could observe how they behaved under certain circumstances. Unburdened by theory they were, as the Heretic would have it, simply observing regularities in the behaviour of the universe.

The Egyptians of four thousand years ago had no knowledge of modern theories of mechanics or gravity, yet they managed to design and construct the pyramids. The Great Pyramid of Giza, one of the seven wonders of the ancient world and built in around 2650 BC is almost 150 metres in height. It remained the tallest building on Earth for nearly 4000 years. Its base is 52,000 square metres in area and it weighs nearly 5.9 million tonnes.

The Romans, again without knowledge of fluid dynamics, built aqueducts to supply water to homes and businesses, to public baths and to sewer systems. Their Nimes aqueduct near Remoulins in the South of France was built in the First Century AD and ran for 50 kilometres down a tiny gradient of only 34cm per kilometre. It carried 200,000 cubic metres of water a day and worked for three hundred years before poor maintenance allowed it to become too clogged [4].

Both steam engines and the internal combustion engine were not generated from scientific principles but from observation and experiment. Steam engines, in the sense of using steam to move something, have been around since the First Century AD when a sphere with two opposing nozzles and containing water was heated up, making it spin about its axis. In 1125 AD in Reims, a church organ was reputed to have been run by stream, and Leonardo da Vinci described a steam powered cannon in the 15th century. Steam engines began to be used for more muscular applications in mines and flour mills by Thomas Savery in 1698 and Newcomen in 1712.

All of this was achieved before any of the laws of thermodynamics had been invented. Engineers did not derive their designs from science but instead designed them using what we might consider nowadays to be non-scientific modes of thought.

The internal combustion engine has been around since 1807 when Nicephore Niepce produced a moss and coal dust-based machine that powered a boat. Basically, an explosion of this material moved a piston. Something similar had probably been originally observed by Huygens in the 17th century when gunpowder was used to move a piston in a cylinder. It was only later in 1824 that Sadi Carnot, often described as the Father of Thermodynamics, in response to these developments produced a thermodynamic theory for idealised engines.

As was usual throughout most of the history of engineering, theory followed design and construction. Theory was a story retrospectively constructed to try to explain how they had managed to do it.

And that's been the case for thousands of years. Engineers leading, theoreticians trailing in their wake spinning explanatory fables, at least according to our Heretic. However, in the last fifty years or so, something profound has been happening in the field of engineering [1]. Before then, engineers rarely learnt from theory but rather from real life practice. These engineers were usually involved in the actual building of the things they designed. This hands-on approach made them intimately aware of the materials and methods used in construction, and so gave them a better awareness of their capabilities but also of their limitations.

By the 1950s and 60s this approach was changing, and engineering became based more on mathematics, in other words theory, and on computers. Design engineers moved further and further away from the things they were designing. Rather than hands-on, they became increasingly hands-off.

As this book has been trying to show, the trouble with all this is that the real world does not comply with our mathematics. It is far more complex and takes absolutely no heed of the approximations and suppositions that are required to reduce reality into glib mathematical forms.

Engineers who make the mistake of regarding their theory-based equations as the truth can come produce catastrophic outcomes, and many such examples will be discussed in this chapter. One can, however, make the general point that nowadays technological constructs usually have to go through many versions and prototypes and models. The reason for this is partly because new features are introduced which do something additional that the old model couldn't, but a very big part of this process is to overcome problems in the design of the first version that were not anticipated by theory. This applies to everything from cars to the notorious 'bug fixes' necessary for computer games and other software.

Indeed, drawing board designs often do not survive exposure to reality, just as battle plans do not survive exposure to the enemy. It is the engineer's job to translate the aspiration of the design to the hardness of reality by making all sorts of accommodations to resolve difficulties unanticipated by theory. It is found in practice that structural analyses, as an example, must be employed with caution and judgement because mathematical analyses are always less complex than actual structures. The same is true in other fields of engineering which involve processes or machines.

Despite this, theory in the teaching of engineering has advanced at the expense of hands-on practice and this has had an unfortunate consequence. Since the Second World War there has been a slow degradation in the status of engineers compared to scientists and this has largely come about because of this shift in perception. Instead of engineering driving theory it is theory that is now perceived to be driving engineering.

Science is slowly taking over and taking credit that it does not deserve. One example of this is the NASA Apollo Moon flights that are regarded as emblematic of the triumph of modern science whereas in fact they were actually great feats of engineering. This engineering experience was laboriously gained from years of testing the structures and functions of the real-world components in engineering laboratories.

This change, and the beginning of the myth of pure science, can be seen explicitly in the words of Vannevar Bush, Dean of Engineering at MIT and Director of the US Office of Scientific Research and Development during the Second World War and also architect of the National Science Foundation. In his book 'Endless Horizons' [5] he states: *Basic research leads to a new knowledge. It provides scientific capital. It creates the fund from which the practical applications of knowledge must be drawn. New products and new processes do not appear full-grown. They are founded on new principles and new conceptions. Which in turn are painstakingly developed by research in the purest realms of science.*

This change has become reflected in the teaching of would-be engineers. For example, there has been a move to exams with single correct answers based on mathematical analyses, rather than having the students deal with the inevitable real-life complexities and ambiguities which bedevil every single engineering project.

And the main weapon of this new breed of scientist-engineer is the computer simulation.

Computer simulation is almost ubiquitous in engineering nowadays. As the simulations evolve to try to replicate the messiness of reality, more and more complexity is added based on the physical laws as we understand them today. Rather than this increasing complexity being a source of concern, it actually seems to increase the confidence of the users, misplaced though that may be.

Not everyone is impressed by the increasing sophistication of these simulations. Eugene S. Ferguson, who was both an engineer and an historian, is our go-to heretic when it comes to comments on modern engineering practice. According to him: 'Computerised illusions of certainty do not reduce the quantity or quality of human judgement required in successful design. To accomplish a design, requires a continuous stream of calculations, judgements and compromises that can only be made by engineers in that kind of design.'[1]

Huge effort has been put into these simulation tools aimed at allowing engineers to more easily design complex systems. These tools take the form of expensive software packages that are sold to engineering firms. The larger and more complex the software packages, the greater the confidence they inspire in their users. Unfortunately, these packages are always based on simplifying assumptions and cannot take into account all possible, and often unanticipated, complex interactions between system components. Users are generally unaware of these underlying assumptions.

They also are prone to other forms of error such as the instabilities caused by sensitivity to initial conditions (as described in the Chapter on Chaos and Prediction).

These problems are compounded because the engineers who lead the design of such software are the ones who have been trained under the new regime where theory trumps hands-on experience. This new breed of engineers can often lack an awareness of the basic, mundane things that can go wrong when any design is translated into reality. And certainly, the programmers who write the code for the software, and who generally are not engineers themselves, cannot be expected to provide the experience to anticipate such adverse eventualities.

A more general problem with this approach has been referred to often in this book but is summed up nicely by the architect Alan Colquohon, [6]. Like our Heretic he is of the view that laws are not found in nature and that instead: 'They are constructs of the human mind; they are models which are valid as long as events do not prove them wrong'.

As a result of all these limitations the new way of engineering thinking has seen a number of high profile, expensive and often fatal failures. As hands-on experience has lost ground at the expense of theory driven engineering, these failures have become more frequent and a lot more expensive.

Perhaps the first real hint of the problems to come can be seen from the early days of architectural simulation in the case of the catastrophic collapse of the roof of the Hartford Coliseum in Connecticut. Roof trusses, in the form of 200 thirty-foot-long, slender cruciform bars were installed having been modelled using software to calculate compression loads. The simulation indicated that these would be sufficient to support the roof. Unfortunately, in January 1978 the roof collapsed after a snow storm. The cruciform bars were subject to forces other than pure compression and this had not been anticipated in the original simulation programs. The snow loaded the 12000-square metre structure asymmetrically, producing lateral as well as compressive forces. [6, 7]

Only a few hours before the collapse, thousands of fans had been inside.

In 1979 there was the first of the Big Three (at the time of writing) publicly acknowledged civil nuclear accidents; Chernobyl and Fukushima Daiichi were to follow in the years to come. This happened when there was a meltdown of a reactor at Three Mile Island in the United States. A pressure relief valve was designed, by someone who had no hands-on engineering experience of 'sticky' valves, to show up on the plant status board as being either open or closed. The complexity of a real-life valve that can sometimes be partly open had been reduced to a nice simple binary on/off condition that the original simulation, and the output display on the Status Board in the actual power plant, could better handle. The Status Board showed the valve 'closed' even though it was partially open, causing an undetected loss of coolant from the reactor over a period of two hours. The reactor eventually overheated and went into meltdown.

These two incidents were nearly fifty years ago. Surely the software will only have got better since then? I'm mentioning them because this was just the start of the problem. There were still engineers around in the 1970s who intimately understood the materials and devices they were working with. As theory continued to supplant hands-on experience and even more faith was put into computer simulation, the failures just got worse and worse. As the simulation software got more complex with time it took the designer who used them further and further away from any dangerous assumptions the software programming team had made. The user became less aware of any critical factors that may have been left out.

The Hubble telescope launched in 1990 had a series of design related problems [8], necessitating four further Space Shuttle visits between 1993 and 2002 to repair them and replace faulty systems. One design related problem was the result of a vibration in the satellite itself caused by its solar panel supports heating and cooling. This made it expand and contract as it moved in and out of the Earth's shadow 'like the slowly flapping wings of a great bird'.

These and other failures on the spacecraft had not been anticipated in any of the design simulations due to the programmers not being able to imagine the realistic conditions the satellite would meet.

The Challenger Space Shuttle disaster illustrates another unfortunate side-effect of the loss of status of 'hands-on' engineers. This comes under the heading of 'top-down Design'. 'Bottom-up design' had been the traditional means of engineering practice and it involved designing, building and fully testing each system component before the full system design had been defined. In this way the true limits of performance were established and any incompatibilities with other system components had a better chance of being identified and the overall system design modified accordingly [9]. The modern 'top-down' approach designs the whole system on a computer, then teams of engineers construct the individual bits simultaneously. Often this doesn't turn out well.

On Jan 28th 1986, a booster seal 'O' ring on the Challenger Space Shuttle had become brittle because of the unusually low temperatures at launch in Florida. Hot gases escaped and caused structural failure of an adjacent external fuel tank. Astronauts Scobee, Smith, Resnik, Onizuka, McNair, Jarvis, and McAuliffe were all killed.

Delays in the launch of a previous mission and poor weather at one of the abort landing sites in Senegal, coupled with one delay due to a malfunctioning micro-switch and a stripped bolt in the shuttle itself, had put pressure on NASA. To make matters worse two later scheduled satellite launches for planetary explorations had unalterable launch dates.

However, the weather at the launch site was unusually poor with temperatures expected to be close to 1o C. Though some engineers from Morton Thiokol, the firm who made the solid rocket boosters, pointed out that the 'O' rings had not been tested to lower than 12o C, they were challenged to provide data to show the system was not safe to launch [10]. Of course, they couldn't.

In the end, the ambient temperature went down to minus 1.7o C.

Following the explosion, the Shuttle was torn apart by aerodynamic forces 73 seconds after take-off. Even so it is possible that some of the astronauts survived until their cabin hit the ocean at 200 mph two minutes and fourteen seconds later.

In top-down Design, the whole system is designed and specified before any testing of components to establish their true limits has even begun. The adverse effects of one component on another have less time to be anticipated and avoided. Indeed, to do this properly would be a vast undertaking for the design and construction of the Space Shuttle that had over two million individual parts. Unfortunately, by taking the quicker and less expensive Top-

Down approach, this is an example of how projects like the Space Shuttle can come to fruition and operation with fundamental design flaws still in place.

Richard Feynman, the famous physicist who we've seen call renormalisation 'a dippy process', was called in as part of a team to investigate the disaster and he had much to say. In terms of bottom-up designs: 'First it is necessary to thoroughly understand the properties and limitations of the materials being used with testing being done presumably on experimental rigs. With this knowledge, larger component parts (such as bearings) are designed and tested individually. As deficiencies and design errors are noted, they are corrected and verified with further testing. Finally, one works up to the final design of the entire engine to the necessary specifications'.

He had this to say about the top-down approach: *The space shuttle main engine [he was talking here about the solid rocket booster] was handled in a different way- top down we might say. The engine was designed and put together all at once with very little studies of materials and components'... 'A further disadvantage of the top-down method is that if an understanding of a fault required a simple fix, such as the shape of a turbine housing, it may be impossible to implement without a redesign of the entire engine'.*

Top-down design was invented by the military and its effects can be seen in many of its new equipment project failures that are almost always mind bogglingly expensive. One can argue these failures come from placing too great a faith in the science from which the concepts for new weapons systems arise. Coupled with this may be the increasing belief that engineering is a mundane profession, there to simply bring the theory-based ideas of weapons science to fruition. Rather than scientific theory coming along later to try to explain what was found in engineering practice, as had been the case for thousands of years, theory was now very much driving that practice.

This modern reliance of the military on science rather than engineering contradicts what little historical evidence we have. The Department of Defence commissioned a study into contributions to the development of new weapons systems. This study was called Hindsight and looked at the period 1958-1966 [11]. Contributions were assessed as follows: technological 91%, applied science 8.7%, pure undirected research 0.3%. In other words, the old fashioned hands-on engineers had done the bulk of the heavy lifting with little in the way of contribution from theory driven scientists. That is definitely not the case now.

The reliance on 'pure science' and top-down design by the US military has led to some spectacular failures. One of the early ones included the F-111 Aardvark Joint (Navy and Air Force) Strike Fighter program from the 1960s that was abandoned when the aircraft could not meet the specifications. The most incredible of these specifications was that variants of this single aircraft were required to be capable of assuming the roles of a strike fighter and strategic bomber, as well as being capable of being used in reconnaissance duties. That's like saying a vehicle should be able to function as a family saloon, a racing car and a dump truck.

Even when top-down driven plans are, more or less, successfully implemented this inevitably requires massive cost overruns. Examples include the development of the Black Hawk helicopters (237% cost overrun [12]) and the Patriot missile (77% cost overrun). In the case of the latter: 'a major reason for the development cost increase was that the original cost estimate did not recognize the level of effort and difficulties associated with developing and producing a hit-to-kill missile compared with those of previous missiles' [13])

More latterly, unexpected cost overruns of 80% have been found in the F-35 multi-role strike fighter. In 2006, the General Accountability Office warned that excessive concurrency between the production of additional F-35 aircraft and testing of their design might result in even more expensive refits for the several hundred aircraft planned to be produced before completion of tests [14]. In other words, they were testing and building at the same time, rather than doing the testing before the building.

The DoD Procurement Chief Frank Kendall, speaking of the F-35 program has said: 'Contracting trends such as "concurrency" -- the overlapping of development, testing and production -- and fixed-priced buying are embraced and rejected in cycles. In the past two decades we have been for-or-against concurrency four or five times, and for-or-against fixed price contracting four or five times. We are now going through another cycle where concurrency has fallen out of favour -- with the F-35 Joint Strike Fighter as the cautionary tale -- and fixed price contracting is back in vogue even though it has resulted in several procurement failures in the past. Thus, the use of concurrency in the Joint Strike Fighter program has caused untold damage in time and money'. Kendall went on to say. 'Putting the F-35 in production years before the first test flight was acquisition malpractice. It should not have been done. But we did it. The JSF management was self-deluded by optimistic predictions that digital simulations could replace actual flight tests. Now we're paying the price. We're finding problems in all three variants of the F-35.' [15]

The failures of this wrong way around approach to design do not just manifest themselves in the form of cost overruns. Unanticipated extra costs, rather than causing cancellation of a program, instead can be manifested by reductions in the anticipated numbers of the units being produced. For example, the F-22 Raptor aircraft programme entering service in 2005 program was originally intended to produce 750 planes, but this had to be reduced to 187 [16].

Cost overruns are also found when it comes to designing other military vehicles. The 'Army's Future Combat Systems' Programme was due to produce eight new major vehicle types but failed to do so, resulting in its cancellation in 2009 despite research and development costs of $200 billion [17]

Naval cancellations include the CG(X) cruiser and the DDG-100 Zumwalt class destroyers (24 were originally envisaged though only three were finally ordered). The Littoral Combat Ship has had a 100% cost overrun.

A $25-billion-dollar Space Radar satellite-based reconnaissance program was cancelled due to major technological costs [18].

In 2010, in an analysis of the US Department of Defence's 96 Major Defence Acquisition Programs which totalled $1.6 trillion, it was found the average cost overrun was 50%, with some double or triple the original cost estimates.

Cost overruns on big projects nowadays, whether they are in engineering or architecture or IT systems, seem to be almost inevitable but that should not be the case. Using theory to drive engineering and design is perhaps the wrong way of going about this and comes with a heavy price in terms of money and human life.

This all hardly helps the poor put-upon engineers who actually have to build things. Not only have they lost status to scientists who, too often, are the ones who hand them these unrealistic top-down designs, but the engineers somehow have to make the damn things safe. And that's quite a worry because of the inevitable flaws in the theories which went into these top-down designs in the first place.

Rather than wring their hands in despair, engineers reach instead for something that has many names such as 'Design factors' or 'Safety Increases' or 'Factors of Safety' or 'Margins of Safety'. When it comes to buildings and structures like bridges, where failure can be very obvious, and so really quite embarrassing, engineers simply increase the theoretically derived levels of expected load to give themselves a wide safety margin. This is partly to take account of the degradation of construction materials with time, but also to try to compensate for factors not taken into account in theory-based design.

These factors are neither set at common standard levels, or even have common definitions, but rather vary from speciality to speciality.

However, let's take as an example a 'factor of safety' of two. This means the structure with such a factor could withstand double the expected load.

These factors vary depending on how critical the failure would be. If deaths or severe expense may be involved, then safety factors of four or higher, sometimes even ten, are used. Where structures or components are less critical then factors of two may be considered sufficient.

Even a factor of two will greatly increase the cost of the project.

More latterly, it may be that these additional factors have served as safety buffers and have partly hidden the inadequacies coming from the greater reliance on theory and design simulation. The concern is that this may have an unintended consequence. Many more disasters than we realise may have been averted because engineers have used these safety margins but this may have resulted in an increased but misplaced confidence in the accuracy of the simulation approach. This may in turn lead to a reduction in these costly additional safety factors, with mayhem ensuing.

Let's now look at a branch of engineering where success seems much less tainted. Perhaps here we'll find an instance of where theory is triumphant.

Electronics

Your iPhone is both a beautifully designed and fabulously effective piece of technology. How could it exist if all our theory was so unrealistic?

Modern electronics are based on the properties of semiconductors. In essence, these are neither electrical conductors nor insulators but can act as either depending on controllable circumstances. The theory behind them is complex but, rather than make this book even longer and strain the reader's patience even further, I will skim lightly over the surface. More in-depth treatment of the theory and the development of semiconductors can be found elsewhere [20].

Michael Faraday back in 1833 was the first to notice the unexpected behaviour of the semiconductor silver nitride. Until then, the electrical resistance of materials had been found to increase with increasing temperature and this was thought to be because of increasing collision processes involving the metal atoms that made up the conductor. The higher the temperature, the more mobile the atoms, the more collisions between them and so the greater the resistance to the free flow of electricity.

However, in silver nitride the resistance decreased with temperature. This and other observations on which semiconductors would later depend (such as the photo-voltage effect observed by Becquerel in 1839, rectification observed in 1874, Hall Effect observed in 1879), only began to be 'explained' by theory in 1931. This theory was produced by A.H. Wilson and postulated something called 'forbidden energy gaps'.

By whatever means semiconductors actually worked, they had the potential to be used as amplifiers. In other words, they would be able to magnify weak signals. There was a great need to be able to do this in those early days in order to allow long distance telephone communications. Metal may easily conduct electricity but it still has some resistance and this reduces the strength of the electrical signal. Amplification up until then had required hot, delicate, expensive and energy consuming triode valves.

After the Second World War, whole research laboratories were being set up to pursue concerted research in a number of fields. Until then research, in either industry or academia, had been on a much smaller scale. These new larger scale research labs were trying to emulate the example of the Manhattan Project which had spawned the first atomic bombs. In the US, at least, it was also a consequence of high level governmental policy urging intense support for pure research in the belief it would lead to major applications. Vannevar Bush, Chairman of the National Defence Research Committee in 1945 as we have already mentioned, was the driving force behind this.

To the Heretic, Vannevar Bush is something of an anti-Christ.

Bell Telephone Laboratories were already ahead of the field and were working with a view, as the name suggests, to supporting research into telephony. It was envisaged that Solid State, in other words semiconductor, amplifiers could be a much cheaper and more effective alternative to the triode-based signal repeater stations used to boost telephone signals that were otherwise attenuated over long distances.

The primary developments that made the applications of semiconductors possible, as in so many other fields of technology, were not in fact a product of the successful application of theory. Instead it was by the observation of unexpected effects not predicted by theory, though theory was, of course, quickly re-jinked to encompass these.

Our Sceptic, a fan of Vannevar Bush, would point out that it was the faith in science, as exemplified by Bush, which made researchers do the research in this concentrated way in the first place and which enabled effects to be serendipitously discovered. So, what if the theory had to be modified to encompass the new observations? That's just the way science progresses.

At Bell Labs, a driven physicist and eugenicist called William Shockley was convinced practical semiconductor amplification would be possible using the so-called field effect transistor approach. This approach failed, prompting the physicist John Bardeen to produce a theory based on something called the 'surface states' to explain why this wasn't working. Basically, this theory said that impurities and imperfections in the semiconductor material were responsible for the effect. Walter Brattain, who devoted his research to these 'surface states', then developed an apparatus which seemed to show the existence of these states. Shockley, Bardeen and Brattain would go on to win the Nobel Prize. Indeed, Bardeen went on to win it another time as well.

In order to study surface states more fully, they envisaged it would be necessary to perform experiments in a low temperature cryostat and a vacuum to prevent any condensation on the conductor which might affect these surface states. That would require a rebuilding of the apparatus and thus a significant delay. That would be bad news in a field in which competition between labs and companies was becoming ferocious. As a shortcut, the Bell team decided to immerse the apparatus in alcohol. Unanticipated by theory, the surface states were neutralised and the way was suddenly open to make effective transistors that are the basis of all our modern electronic devices. As John Orton in his history of semiconductors says: *'This was serendipity of a high order. A chance observation from what was in many ways, an ill-advised experimental shortcut!'* [20].

Needless to say, theory was quickly modified to explain why this had happened.

Further examples of critical breakthroughs which were seminal to the successful use of semiconductors but which weren't predicted by theory include the discovery of the inversion effect. Careless washing of samples inadvertently dissolved an important oxide film on a semiconductor thus making two, rather than one, electrical contacts on the film. This increased the range of voltage application frequencies by a thousand-fold. This huge improvement was the second unexpected discovery that paved the way for the birth of the first transistor on December 24th 1947.

Other seminal but unexpected findings include the construction of the first p-n junction due to impurity segregation in a molten silicon rod and the discovery of planar technology in 1955 due to water contamination of a diffusion capsule. This made it clear that silicon rather than germanium was better in solid state circuits.

OK, so the development of semiconductors and hence transistors may have been by chance and careful observation rather than predictive theory, but what about the subsequent and absolutely amazing miniaturisation that made devices like the iPhone possible? Surely that's got to be a triumph of science.

Not so. The miniaturisation of these circuits that are now so small they have to be etched on the silicon by ultraviolet light and even, more latterly, by x-rays is a triumph of dogged engineering rather than science. Designs are based on engineering observations of the properties of materials. The subsequent theoretical overlay to explain these observations are based on esoteric and arcane concepts like forbidden energy gaps and the migration of not only electrons but also supposedly of the holes they leave behind. The main developments in semiconductors have been made without the post-event rationalisation that is theory. Not only that, but the most seminal developments were made despite the theories prevalent at that time.

We'll end by looking at two of our most important technologies that are generally held to have come about due to predictions made by scientific theory and see if that really is the case.

Radio

Maxwell is often quoted as predicting the existence of radio waves but it is nowhere near as clear cut as that. His equations allowed for the existence of electromagnetic radiation beyond the visible frequencies that people had been aware of up until that time. As has already been mentioned, his theory did so based on an absolute reference frame and a substance called aether that filled space, yet modern physics denies the existence of both.

Indeed, rather than Maxwell's theory, the main stimulus for looking at these lower frequency waves came from the chance observation by Hertz in 1886 that if an open circuit of a rectangular bit of copper wire (one with an air gap) was placed near a circuit across which a spark was being induced, that the first circuit could also be induced to produce a spark across its own air gap. It wasn't true for all open circuits but only those of specific sizes.

Ironically D.E. Hughes had observed a similar phenomenon seven years earlier when a spark discharge was picked up 500 metres away by a microphone contact. Thinking this was evidence of transmission of previously unknown electrical waves, he presented this to the President of the Royal Society and the Chief Electrician to the Post Office. Using existing, though erroneous (in the modern view) theory, these gentlemen were able to come up with an alternative explanation, long delaying Hughes publication of his results [21].

Awareness however grew of electromagnetic fields other than light and this led to further experimentation.

Marconi in a sense didn't 'invent' radio but rather he systematically improved existing technology. He also essentially ignored prevalent theory at that time. This supposed that radio waves would act like light and that transmitter and receiver would therefore have to be in line of sight. A self-taught man, he was less influenced by such theoretical considerations as his prevailing theory-hobbled contemporaries and that is perhaps the reason he succeeded where they failed.

Lasers

Were lasers really predicted by theory? Many say they were so it's an issue this book has to address, though the history is rather murky. In order to talk about this, I will need to go into more detail than the non-scientifically trained general reader may be comfortable with. Sorry about that, but the reader can skip this section without affecting their understanding of what follows in the rest of the book.

Lasers have made a huge impact on mankind and have become ubiquitous. They are now the basis of high volume telecommunications and form the backbone of the internet. All optical disks use lasers for recording and reading of data. Lasers are used in precision cutting in metal, ceramics, plastics and wood. They are used in surgery to destroy tumours or reattach detached retinas. They are used in bar code readers in supermarkets, for accurate alignments of components during assembly of machines, and for accurate measurements of length [2].

The basis of lasers (light amplification by the stimulated emission of radiation) is the phenomena of stimulated emission. This came from a 1915 paper by Einstein entitled Strahlungs-Emission und -Absorption nach der Quantentheorie. His ideas combined classical laws of physics with the new quantum theory. He examined energy changes in a gas being bombarded by radiation and postulated three processes of energy change:

a) The gas molecules' energy states were quantised (in other words they could take only certain discrete values). If the incident radiation had the same energy as the difference in the energy levels of the molecule, then the photon of energy was absorbed and this kicked the molecule into the higher energy state.

b) Most of the molecules will be in the low energy state but some will, quite naturally, be in the higher energy state. Molecules in this higher state can transit to the lower state by emitting a photon of radiation. This was called emission.

c) The third state applies when molecules in the higher energy state are struck by a photon (again with the same energy as the difference in energy levels) and this will produce stimulated emission. The original photon will continue on its way but will be joined by another photon caused by the energy transition to the lower state. In other words, one photon produces two. It was later shown by Bothe in 1923 that the newly produced photon travels in the same direction as the incident photon. In other words, a form of amplification is possible.

This proposed third mechanism was a new and purely theoretical concept but excited little interest at the time.

One interesting and illustrative corollary of this new quantum mechanical way of thinking came from the physicist Niels Bohr and his abandonment of the classical theory of light dispersion. We'll not go into any details but the important point is that the classical model wasn't abandoned because it was shown not to work. In fact, it did still work, giving a believable explanation of the dispersion and absorption of light. It also made some predictions which could be verified experimentally (specifically it showed the dependence of refractive index on wavelength varied in a non-linear way).

The reason this theory was abandoned was because it just didn't fit in with the new quantum world envisaged by physicists like Bohr. If Bohr's view was correct, then the theory suddenly lost all theoretical justification even though it appeared to work.

This is an example of where many theories may appear to work, but where the differential choice between them is made on the basis of the scientific vogue of the time.

After the classical approach was abandoned, it took the quantum mechanical theorists like P. Debye, A. Sommerfeld and CJ Davisson, several false starts to replace it. More 'correct' theories, in other words ones that complied with quantum mechanics, were brought in by R. Landburg and F. Reiche, though the effect of Einstein's predicted stimulated emission still did not appear out of the maths. Further modifications by Kramers in 1924 managed to bring it back in. However, in order to do this, he had to avoid considering the real orbit of the electron and instead had to substitute the atom for a set of hypothetical oscillators in his equations. Lande called these fictitious oscillators the 'virtual orchestra'.

In other words, a theory which worked (the classical one) was rejected and a new theory was substituted which didn't produce an expected phenomenon until a fictitious concept was introduced to make it fit experimental facts.

Why should the latter theory be better than the former just because it ties in with sexy new ideas like quantum mechanics?

Either way, lasers were a long time coming. J. Weber suggested in 1952 how the amplification inherent in stimulated emission might be utilised to amplify microwaves to produce the MASER (microwave amplification by stimulated emission of radiation, though in the early days detractors used the alternative acronym Means of Acquiring Support for Expensive Research). Weber's own ideas of how to do this never worked and it was left to C.H. Townes to produce the first Maser.

So, the view that lasers were predicted by theory is far from clear cut but it does offer some support to our Sceptic. Einstein did predict stimulated emission though it took quite a bit of fiddling, and the construction of another mythical beast ('the virtual orchestra'), to fit into general quantum theory.

The Heretic, however, draws parallels with the historical development of the theory of Radioactivity. There are always many alternative theories at any time and some, by chance, will produce a prediction which will subsequently be shown by experiment. It does not, however, mean the theory is correct.

Engineering is mired in the down and dirty nature of reality which often makes nonsense out of the predictions from scientific theory. Engineers generally perform their heroic deeds despite, and not because of, the theories of science.

So far in this book we've mainly concerned ourselves with harder sciences like physics and have tried to show all the problems that messy reality throws up to undermine theory. Now, however, we're going to look at the 'softer' and therefore more intrinsically difficult end of science and delve into the world of medicine and its even messier blood and guts reality.

10

Medicine and the Body

Life is short, the art is long, opportunity fleeting, experience fallacious, judgement difficult.
Hippocrates' first aphorism regarding the practice of medicine

Trust not the physician/ His antidotes are poisons and he slays...
Shakespeare's Timon of Athens

On the whole, more harm than good is done by medication.
Oliver Wendell Holmes (the famous doctor, not the famous jurist) in 1860

The human body and its functions are extremely complex and medicine is always going to face an uphill battle to fix it when it goes wrong. The difficulties medicine faces can be overwhelming and this is reflected in its development which has been a catalogue of fallacious reasoning and opportunities lost. Taken over its several thousand years of history we will see how medicine has killed far more than it has ever really saved. These deaths arose from a slavish devotion to prevailing doctrine and theory.

This killing more than curing situation has probably only reversed within the last hundred years or so as medicine has definitely got more effective. The question is now whether this improvement will continue. Will it just be a matter of making minor adjustments to our present theories of medicine as we zero in on the truth? Will mankind one day be the master rather than the victim of disease?

Both the Believer and the Sceptic believe the answer to all these questions is a firm 'yes'. In this chapter we'll try to assess how realistic this faith is.

We're going to start this chapter with a short recap of medical history and the science that was thought to underpin it. This is necessary to gain some perspective on current medical theory because throughout medicine's history, and despite its often manifest failings, at any given time its practitioners have thought they understood the principles of medicine clearly. This has enabled them always to speak with great authority even though, according to present day perspectives, they were greatly in error almost all of the time. By gaining some perspective on the history of medicine we can begin to address the following question: if medical theory was so wildly wrong in the past, how right can it be today? We'll look at some of modern medicine's undoubted successes and we will see how these were based on observation and experience rather than being driven by theory. Where modern theory is used, based as it is on our limited understanding of the complex and mutually interacting processes of the body, we'll see just how unexpected and often harmful the results can be.

Disease, in all its manifestations, has been the bane of mankind from time immemorial. Perhaps the first evidence of the possible use of medicine comes from the 5200-year-old Iceman discovered preserved in ice by climbers in the Alps. A long-delayed post mortem revealed the man had been suffering from Whipworm in his gastrointestinal tract and it is thought the charcoal he carried was to aid his deranged digestion. Stress markings on his fingernails indicated repeated infections and it has been noted that the two types of mushrooms he carried, thought originally to be used as food, actually have antibiotic properties.

Faced with a wealth of diseases and the complexity of the workings of even the normal human body, doctors, in whatever form, have resorted to simple conceptual systems to try and explain it all. Inevitably these have all ultimately failed though some have taken a very long time to do so.

For example, ancient Chinese medicine in trying to understand the complexity of the symptoms of malaria, not least the alternating hot and cold flushes, put them down to three demons. The first used a hammer to cause a pounding headache, the second used a pail of iced water to give the patient chills, and the third carried a portable stove to produce fever [1].

More generally, the Chinese thought they needed only five elemental components to explain health: fire, earth, metal, water and wood. Diseases came from imbalances in these. The body was also subject to the balancing forces of yin (heat, positive energy) and yang (cold, negative energy).

Probably the first people to look for a rational, rather than a magical or religious cause of disease, were Hippocrates and his followers on the Greek island of Cos in 400 BC. Later on, in 200 AD the Greek Dioscorides wrote De Materia Medica which described 800 drug sources from minerals, plants and animals.

The ideas of Hippocrates and later of Dioscoridies and a Roman physician called Galen essentially became the basics of medical practice until the 19th century. Hippocrates' text Methodo Medendi was published unchanged until the 16th century. Indeed, the Hippocratic Oath is still with us today.

And it was all, in the view of modern medicine, complete nonsense.

Hippocrates believed that if the four elementals or key fluids (also called humours) were in balance, the body was healthy. These were air, water, fire and earth. Disease came from an imbalance of these humours. Blood was considered to be produced by the air humour as it was both hot and moist. Phlegm was from the water elemental, yellow bile from fire and black bile from the earth.

Though Hippocratic beliefs were to prevail for two thousand years, there was some scepticism almost right from the start. According to Celsus in AD 40 there were actually at the time three schools of medical practice. The 'dogmatists' believed diseases had hidden causes which should be looked for; the 'methodists' said disease was basically mechanical and a matter of particles moving too quickly or too slowly in the body. Finally, the 'empirics', a more heretical bunch, did not accept any theories of disease but instead advocated using treatments and practises which experience had shown actually worked, at least on some occasions. These schisms found their way into later Roman and Arabic medicine, though the dogmatist approach, through its main advocate Galen, was to be triumphant and remain so to this day.

So, Hippocrates' belief system prevailed but its practice was not for the faint hearted. It bred courageous physicians. Taste, smell and texture were widely employed. For example, practitioners tasted the patient's ear wax; if it was wet then death was on the way; if it was bitter then things were just fine. 21st century doctors rarely do this now. Pus was graded by appearance but also by smell; the fouler it smelled, the worse the news. Blood and urine were judged by how they felt as well as how they looked and smelled [2].

This belief in the importance of things like taste led to some strange but longstanding beliefs. Galen rejected the idea of the heart being a muscle because it didn't taste like flesh when cooked.

Interestingly, the word hypochondriac that nowadays means someone neurotically concerned about their supposed ill health came from the Greek use of touch when feeling the hypochondrium: the sides of the abdomen under the ribs. This was a serious business for Greek medics because hard to the touch meant death was coming, soft meant protracted disease. Only in the 19th century did medicine lose interest in the hypochondrium. Perhaps in a form of blame transfer, rather than a hypochondriac being someone with a diseased hypochondrium according to the unsubstantiated beliefs of the medics at the time, it was now held to be someone who held unsubstantiated beliefs in their own ill-health.

These sorts of Hippocratic beliefs held sway for nearly two thousand years.

It isn't until the 19th century that we see the earliest attempt at evidence rather than belief-based medicine. This new way of looking at things quickly found, for example, that using Hippocratic methods for dealing with lung infections actually increased death rates by nearly 70% [2].

During the reign of Hippocratic medicine most diseases had remained a mystery, their transmission often blamed on 'vapours' and 'miasmas' (malaria comes from the phrase mal aria, meaning bad air). It wasn't until the invention of the microscope in 1590 that real insight came, opening up as it did a hidden world and showing the existence of microbes not resolvable by the human eye alone. Some practitioners began to suspect that at least some disease was due to these 'germs'.

The microscope also put the lie to other widely held beliefs, such as that the organs of the body were made of congealed blood.

Despite these revelations the use of the microscope fell out of favour for over a hundred years. This seems strange to our modern sensibilities but it is illustrative of how science always works within the philosophical framework of its time. Thomas Locke in his highly influential 1689 Essay Concerning Human Understanding explicitly stated that, even if our eyes had the resolution of a microscope, the knowledge we gained would be useless. God, he said, gave us the resolution our eyes needed and no more. Any effort put into microscopy was therefore 'lost labour'[2]. Locke was far from alone in taking this view. In the face of sustained attacks even Marcello Malphigi, considered the father of anatomical microscopy, backed down and said microscopy was nothing to do with medicine. Microscopy was subsequently ignored until the 1850s when the acceptance of the germ theory of disease became more widespread.

Meanwhile the treatments prescribed by Hippocratic medicine could most politely be described as 'bracing'. Blood-letting, a favourite since Galen in 400AD, remained a very common practice until the 19th century as was the aggressive emptying of the bowel and stomach.

Bloodletting is now believed to be harmful in all but a very few diseases. If you are weak and run down from the effects of disease, the last thing you need is to have some of your blood syphoned off. Bloodletting was used for almost all diseases, and indeed was used to treat people who had already bled heavily from wounds or were about to undergo amputation.

It is thought that George Washington died due to the 'heroic bloodletting' of his doctor Benjamin Rush in 1799. Indeed, haemorrhoids and nosebleeds were considered good things and if you didn't have them you needed prophylactic bleeding by other means such as the application of leeches.

The idea that bloodletting might be very generally efficacious came from Hippocrates himself who believed that menstruation was a means for women to rid themselves of their bad humours. It wasn't until the middle of the 19th century that physicians began to question whether this bleeding technique they had been using for two thousand years had actually been doing more harm than good.

Purging, the emptying of the stomach or bowels by inducing vomiting and diarrhoea, was also a common means of therapy and again could be fatal in patients who were in a weakened condition.

Indeed, effective medical treatments up until the 20th century were rare and essentially consisted of opiates for pain relief, quinine for malaria, fox glove (digitalis) and amyl nitrate for the heart and vessels, and colchicyum for gout [3]. Aspirin only arrived in the West in 1896.

With a dearth of effective treatments for most diseases, quackery was rampant. Mud bathing and electrified beds were used for sexual rejuvenation. Homeopathy was commonly used as was hydrotherapy where exposure to spa waters was thought to drive out disease. Vegetable based therapies were often applied to induce vomiting and sweating.

Good physicians of the time might have been aware of their own limits and have known there was little they could do for so many diseases but, in the face of insistent and desperate patients, it was natural for many practitioners to reach for the opiates and chloral hydrate, a sleeping potion. At least these would reduce the pain though often at the expense of later addiction.

Doctors in the 19th century began to sport new diagnostic tools which also served at least partly as shamanic fetishes, such as the stethoscope which was invented in 1816. Previous to this, as indicated by Oliver Wendell Holmes in 1882: 'The lancet was the magician's wand of the dark ages of medicine' (the lancet is a needle designed to draw blood). The ophthalmoscope and laryngoscope made appearances in around 1850, and the thermometer became compact enough to carry in the 1860s. Blood pressure was first able to be measured using the impressively Heath Robinson-like sphygmanometer in the 1880s.

These devices might all help to show that something was really wrong, though there was still little the doctor could do in the way of effective treatment.

The body itself

The journey deep into human flesh initiated by dissections is what has made Western Medicine unique. It has sustained the fruitful conviction that in ever-more-minute investigation of the flesh lies the key to health and disease, even if it has also encouraged a tendency to myopic reductionism, to miss the whole by concentrating exclusively upon the parts. Roy Porter, medical historian [3].

It is surprising how little medicine knew about what the insides of the body actually looked like until the 1300s.

Though people had been hacking each other to death from time immemorial, human dissection for the sake of knowledge seems to have been rare though there were some reports of this happening in about 300 BC. Public dissections really only began in earnest in the 1300s in Bologna but were common there for the next several hundred years.

Vesalius is sometimes called the Father of Modern Medicine though we will see others who have some claim to that title. He's also one of the very first body snatchers for medical purposes. In 1536, while on a walk, he came across the body of an executed criminal on a gibbet. The body had decayed badly and Vesalius became fascinated by what he could see of the skeleton; so much so that he immediately stole the arms and legs, coming back again to make off with rest under cover of darkness.

Later, boiling off the remaining flesh to leave just the bones (I've done this myself to get at some lumbar vertebrae for bone mineral studies. The broth smelled very much like chicken soup, something I have not been able to eat since). With the bones he reconstructed the skeleton. Bits of cartilage too fine to re-attach to the skeleton, such as those in the nose and eyelids, he strung together to make a fetching little necklace which he hung around the skeleton's neck. Quite a few of the Fathers of Modern Medicine were a bit strange, as we will see.

You had to get your fun where you could in the 16th century and it appears that Vesalius was quite a card. Illustrations for his subsequent books on dissection show putti (cherubim) dissecting and boiling the bodies to get the bones out. In the section on the human backside the illustration shows a putti evacuating his bowels on the large illustrated first letter of the chapter.

Dissection became something of a habit for Vesalius though he wisely took precautions because at the time messing about with dead bodies had an even greater stigma attached to it than it has today. He had the bodies quickly flayed so that they were unrecognisable to wrathful friends or relatives.

He was also a great vivisectionist. Cutting up pregnant dogs allowed him to watch the unborn pups struggling for breath as the maternal blood supply was cut off. Vesalius even admitted to dissecting out a still beating human heart; the patient/victim apparently having just been unfortunately killed in an 'accident'. According to the surgeon Ambroise Pare, the Father of Modern Medicine later went on a pilgrimage/fled to the Holy Land after an unfortunate dissection incident on a young girl who turned out to be still alive [2].

At least Vesalius' work served to weaken the shibboleth around vivisection and dissection. Public dissections, usually only performed in the winter months to reduce the rate of putrefaction, started to be held. Such dissections put to bed many ancient misconceptions, for example that as well as a duct connecting the penis with the testicles there was also one connecting the brain with the penis. Supposedly this had been to allow spirit from the brain to mix with the secular matter from the testicles.

In Hippocrates time, and for two thousand years beyond, it was thought that wombs wandered around women's body from the head to the big toe. When they were in the chest there was danger of suffocation.

In the Renaissance these ideas began to change. During dissections in 1561, Gabriele Falloppio discovered the vagina, though pedants might claim he only came up with the name.

Indeed, until Regnier de Graaf's 1672 book On Women's Organs it had been assumed for two thousand years that men and women had the same sex organs. The visual differences came about quite simply because men's organs, produced in more favourable 'hotter' wombs, were more developed than women's. The latter's were just folded up inside rather than hanging down. However, de Graaf showed that the structure of the male testicles had an entirely different structure from the female testicles that, nowadays, we call the ovaries.

The truly wonderful advance that dissections allowed, namely the ability to visually compare healthy and diseased organs and actually see how differ-ent they could be, was the start of a turning away from the idea of 'humours' which had held sway until then.

As well as showing structure, dissection also yielded insights into how organs and systems in the body functioned. This field of physiology, origi-nally known as 'animal economy', was advanced by William Harvey, another Father of Medicine, in the early 17th Century who discovered the circulation of the blood. Until then it had been assumed to be more or less stationary and indeed it was the idea of subsequent 'stagnation' of the blood that had led to the idea of bleeding to alleviate or cure disease.

As with other Fathers of Medicine, Harvey was a bit of a rum character. He was an anatomist who had performed dissections on the bodies of his own father and brother.

Around this time there was another major development that continues to shape medicine even today. Under the aegis of the Royal Society there was a mingling of physicians and early physicists such as Hooke and Boyle. From this early 'networking' a much more mechanistic view of the human body developed. This view was boosted in Italy by people such as Borelli and Bagli-vi. The latter proclaimed: *'A human body...is truly nothing else but a complex of chymico-mechanical motions, depending on such principles as are purely mathematical'.*

Though people might now have begun to believe the body was essentially mechanistic, few questioned in those religious times the existence of the soul, though there was doubt as to where it resided. For thousands of years it was thought to be in the heart, though in the 18th Century William Cullen first suggested life itself was a function of the nervous system and whatever powered that. Others, like John Hunter, claimed the life-force lay in the blood.

Despite the burgeoning anatomical knowledge, there will still some very strange ideas around. Even until the 17th century, it was believed that mice could be spontaneously generated from grain, flies from dung and worms from fruit. This was because at the time there was no clear distinction between animate and inanimate objects. Rather than understanding that an animal had been attracted from elsewhere to a food source, people believed it had grown from it like mushrooms from a rotten tree trunk. This dogma was strongly held and it took over a century and a half from the first suggestion that spontaneous generation was nonsense (Swammerdam in the 17th Century) until it was widely considered to be false.

This new mechanistic view of the body drew a sharp distinction between the normal and diseased states, and indeed this idea is entrenched to the present day. However, even two hundred years ago, voices were raised in warning. FJV Broussais said the fixation on localised diseased anatomy was too simplistic and that there was a continuum linking health and disease; that sickness was often a product of functions, and not just the anatomy becoming deranged.

The fetishisation of pathological anatomy found its greatest proponent in Carl von Rokitanski of Vienna who was said to have performed over 60,000 post mortems. That's about the same number of meals a sixty-year-old person will have eaten in their lifetime. His success paved the way for Vienna to become, and remain, a centre of world medicine.

The final nail in the coffin of the view that human bodies were anything other than mechanical and chemical devices came in 1828 when Friedrich Wohler showed that urea, found in urine, could be constructed artificially from non-biological substances.

Since then there has been a steady shift away from the idea of humans as complex life force enhanced beings to something little better than nuts and bolts.

Medical practitioners faced with dreadful diseases and the terrible uncertainties that come with them have, have always looked to science as a sort of crutch, erroneous though that science may be. In the 19th century science began to place greater emphasis on objective rather than subjective evidence and in turn physicians began to move away from symptoms to signs. The problem with symptoms was that they were based on what the patient actually reported feeling or experiencing and these were far too variable. This is because severity of reported symptoms is to a large part subjective, with one person's mild ache being another person's agony. However, 'signs' were considered more definitively objective measures: for example, the sounds of the heart heard by the stethoscope. The stethoscope was the MRI of its day and is even today still often hung around the necks of junior medics as a totemic symbol of the science that backs up their prognostications, whether they ever need to actually use the instrument or not.

Coupled with the subjectivity when it came to reporting their symptoms, patients were also a bit of a pest because their responses to disease and treatment were so variable. Pierre Louis in the 1830s in Paris was one of the first to try to cut through this inherent complexity by introducing the first ever clinical trials.

Dogmatically held but false beliefs in the theories of medicine can kill in many other less direct ways than 'healing' an already anaemic patient by bleeding them to death.

Scurvy was a scourge of international naval voyages. Initial symptoms included lethargy and spots, spongy gums and bleeding from the nose, eyelids, ear, mouth and anus, but could then progress to jaundice and death. In 1740, on a fleet sent to the Pacific, 1800 men died from scurvy with only 200 surviving.

It has been estimated that perhaps as many as two million sailors died of scurvy [2]. That limes and lemons may prevent scurvy was evident even as early as 1601 on British ships. This came from the observation of the health of sailors on ships taking lemon juice to the East Indies. However, this observation did not fit in with the old medical idea of 'humours' and so was ignored. Indeed, even in 1740, after a request from the Admiralty to the Royal College of Physicians, the advice was to use vinegar which is actually useless in preventing scurvy. It took until 1753 and the publication of 'A Treatise of the Scurvy' by James Lind, that at last some real data got through. Lind had performed one of the first ever clinical trials on seamen on board the HMS Salisbury using the novel concept of control groups. Sailors developing scurvy were given either the nostrums of the day like vinegar or they were given lemons and oranges. Only the latter group got much better.

Despite this compelling evidence, Lind felt it necessary to spend nearly four hundred pages of his book trying to fit these findings in with a modified humoural theory so that it would be acceptable to other physicians. This rather desperate theory was that damp air blocked the pores so the sailors couldn't sweat (I served in the Merchant Marine in my 'gap year' and this doesn't reflect my own experience. Not by a country mile). Lemons, somehow, unblocked the pores.

Unfortunately, when Lind came ashore and ran a naval hospital he concentrated the fruit juice by boiling it, destroying the vitamin C which was the cure, and so even he lost faith in his own results and went back to another old friend to cure scurvy: bleeding. It wasn't until the work of Gilbert Blane in 1793, based on his own observations, that the efficacy of limes and lemons were made evident.

Doctors' adherence to Hippocratic Medicine also delayed understanding and effective treatment of cholera; one outbreak in England in 1849 killed 53,000 people alone. Cholera had been put down to the old concept of bad air (miasma). Ironically, concerns that the disease was spread by vapours from putrefying matter led sewers to be constructed to take these vapours away from London's inhabitants. Unfortunately, the sewers were emptied directly into the Thames from which two thirds of London's drinking water came. Cholera is water borne and spreads by ingestion so this measure struck down even more people.

Interestingly, there had been at the time some evidence presented to support the miasmic view of the spread of cholera by William Farr, later to become superintendent of the Statistical Department of the General Registry Office. He showed an apparent correlation between choleric death rates and the height above sea level where the victim lived. His formula showed a doubling of death rate when height above sea level was halved.

Farr's statistics were probably right but the perspective (Hippocratic) from which he viewed the results was wrong. It wasn't mists and fogs around rivers and lakes and seas that were the problem, but that people living in low lying houses often drew water straight from the faeces polluted water sources.

Bearing in mind that physicians new so little and had so few effective medicines and were routinely killing more patients with their bleeding and purging than they cured, one wonders how they managed to get away with it, often charging high fees into the bargain.

One reason is that for thousands of years and indeed right up until the much better general health of the Western population after the Second World War, patients and their families were often inured to death. This is perhaps why they were more apparently forgiving of the ineffectiveness of their doctors. Bedside manner was probably more important than anything else.

There may be a number of other reasons why medicine still enjoyed success, if not actual effectiveness. Firstly, it might have looked like it was more successful than it actually was because most diseases just get better by themselves. The placebo effect, of which more in the next chapter, is another friend of the doctor and springs from the way that some people get better when given treatment, any treatment, even if it is just a sugar pill. It is perhaps responsible, on average, for a third of the good done by even modern medicine [2].

One other reason for medicine's pseudo success in those days was that physicians had a tendency to think of the patients and not their diseases. That is much less the case nowadays. Modern patients rightly get quite tetchy when they overhear their doctors referring to them as 'the gastric reflux in Ward 4'.

Surgery

Specialisation in medicine was there almost from the outset. In ancient Egypt there was even one enema-toting healer who had the title: Keeper of the Royal Rectum.

Though saddled with the belief that all health and disease was associated with the simplistic four humours, even in the West medical specialisation started very early on. As early as the 12th Century, the most marked example of this was the separation of physicians from surgeons. The latter generally got a raw deal, particularly in northern Europe where surgery was excluded from the medical curriculum. Indeed, surgery was directly associated with barbering [3]. The Worshipful Company of Barbers in the City of London dates back from 1308 and included in its remit: boil lancing, enemas, leeching and the extraction of teeth. It was not until 1368 that surgeons were able to form their own guild within the Worshipful Company. However, it took nearly two hundred more years and an Act of Parliament in 1540 to dictate a clear delineation: barbers were not allowed to do surgery, and surgeons were not allowed to do barbering. Both, however, could still extract teeth. The traditional red and white striped barber's pole reflects the entwining of barbering (white) and surgery (red to signify blood). It wasn't until 1800 that an entirely separate Royal College of Surgeons was formed.

Surgery, in one form or another, has been around for a very long time. Drilling a hole in someone's skull is called trepanning and may have started as early as 5000 BC, perhaps to cure extreme headache by letting out the devils according to the theory of the time.

Bone setting was practised three or four thousand years ago in Egypt where there were also surgical procedures for treating abscesses and superficial tumours. Operations on readily accessible structures like eyes and noses might also be countenanced, but generally ancient world surgeons went no deeper into the body than the tonsils or, as a last resort, to reach the foetus to aid healthy delivery.

Surgery, however, became much more invasive in the 1500s despite huge problems with stemming subsequent haemorrhage and also from the threat of wound infection. Surgeons even then still followed the old notions of Greek medicine which held that suppuration of wounds was a necessary part of the healing process. Little was therefore done to prevent infection, whether from wounds or from the surgery itself, for well over a thousand years and with the subsequent loss of countless lives.

Hot oil cautery of the stumps left after amputation of arms or legs made an appearance in the Renaissance. This at least could help prevent wound suppuration if you didn't die from the shock and pain. It was only in the 1840s that anaesthesia started to be applied, thus making amputations, the removal of kidney and bladder stones and Caesarean sections, all previously attempted without anaesthetic, a little more survivable.

A barber-surgeon called Ambroise Pare developed turpentine and egg yolk treatment in 1545 that had some success in reducing wound infection. He also developed the idea of tying up vessels to reduce bleeding after amputations. This replaced the old standby of hot oil, though in this situation it was being used to melt the bleeding vessels shut.

Inevitably, many false paths were taken. One was the 'wound salve' designed specifically to heal wounds from rapiers. This included in its ingredients: pigs' brains, powder of embalmed corpses and earth worms. In order to make sure it worked, the salve was applied not just to the wound but to the rapier that caused it as well.

However, still working against the few actual improvements was the continuing medical belief in 'humours' with most of the barber-surgeon's work involved bloodletting. After surgery, itself usually very messy and involving the loss of a lot of blood, what could be better than to have even more blood leeched or lanceted away from your weakened body?

The main breakthrough in terms of reducing mortality following surgery came with the concept of the need for cleanliness. It was first made by Ignaz Semmelweis in 1848 when he noticed that fatal infections in obstetrical clinics were much worse when they were run by surgeons than when they were run by midwives. He realised that the surgeons were coming straight from performing autopsies to the clinics and may therefore be spreading infection. He advocated the washing of surgeons' hands and the sterilisation of instruments between patients.

As we will see time and again, new ideas are often considered as heresy in medicine, and indeed science generally. The idea that medics were themselves the cause of so much death was considered almost blasphemous to a profession easing its way into god-like status. Semmelweis' ideas were also contrary to the orthodox views of the time which held that mists from soil were the cause of infections. Semmelweis was hounded out of Vienna and, scandalised to the point of madness by his treatment at the hands of his colleagues, he died in a lunatic asylum.

It wasn't until Joseph Lister's work in the Royal Infirmary in Glasgow that it was seriously proposed that pus from a wound may actually not be a good sign. Written up in the Lancet in 1867, Lister's paper claimed that 'laudable pus' in fact wasn't that at all and that it was germs, not miasma, that caused infections. Lister also developed carbolic acid as the first really effective antiseptic.

Before anaesthesia, surgeons performing amputations had to be quick and ruthless despite the screams and entreaties and the physical resistance of their patients; the longer the operation, the greater the amount of physical shock and so the greater the chance of death. Though the anaesthetic effects of nitrous oxide (laughing gas) were widely known as early as 1795, with the gas even being used as a fairground amusement, surgeons weren't keen to use it because it circumvented the lucrative need for their remorselessly speedy cutting skills. It was in fact dentists who first introduced anaesthesia clinically. The innovator was the dentist Horace Wells who was then hounded to suicide by the enraged medical profession. It took until the mid-19th century before nitrous began to be used surgically.

With decent anaesthesia and antiseptic techniques coming to hand in the 20th century, surgeons went from 'fearing to tread' in the inner spaces of the body, to positively rushing in at the drop of a hat. Even the brain, whose treatment until then had been limited to simple trepanning to reduce intracranial pressure, became accessible.

Though surgery was now often becoming very effective, the new gung-ho attitude led to many excesses. In the UK some advocated the removal of yards of gut to cure constipation. Cutting out or destroying bits of the brain as a means of curing mental illness became something of a vogue. One technique included injecting pure alcohol into surgically exposed parts of the brain to destroy brain tissue. Less gentle techniques included cutting in via the roof of the eye socket and then using a mallet to bang in a chisel-like implement that could be twisted around to sever nerve tracts. Other surgeons simply cut out whole lobes of the brain. At its peak in the 1950s over 5000 such 'lobotomy' operations were performed in the US per year, despite around a three percent mortality rate as well as there being common but devastating side effects such as 'surgically induced childhood'.

And to top it all, like treatments based on 'humours', these treatments too weren't much use at all.

Blood transfusion and the development of antibiotics helped surgery go from strength to strength. As well as cutting things out of the body, surgeons now began to put in implants like artificial hips. They also began to replace diseased organs with healthy ones from donors, and this became much more successful with the development of immunosuppressant drugs that helped overcome the body's natural defence mechanisms trying to reject the alien object.

Modern surgery is a demonstrable success in many ways. Perhaps this is because it owes more to practice than theory. The most effective surgeries are essentially to do with unblocking or repairing pipes and fixing leaks in the blood vessels or in the gastrointestinal tract. Although far more sophisticated, this does have analogies to the work of plumbers who also can do their work very successfully without much recourse to theory. Much of orthopaedic surgery is to do with repairing broken bones or replacing worn out joints and is in some ways analogous to carpentry and car mechanics where effective techniques have been developed by practice and observation more than by theory.

Compounding the general ineffectiveness of Hippocratic medicine over the last few thousand years, there have also been two major changes underway that have had huge and often adverse impacts on human health: the growth of cities and of the agriculture required to sustain them.

The growth of top-of-the-food-chain health problems

Many diseases in epidemic form are the product of man himself. He congregates in cities and makes the spread of infectious diseases like smallpox, measles and flu far easier. Concentrations of people also make large depositories of filth that can harbour disease. Large, communal but polluted water sources bring cholera and polio, typhoid and hepatitis.

To sustain large conglomerations of people, granaries became necessary. Vermin feeding on this grain facilitated disease transmission through their excrement which was then ingested by humans. Rats in particular are held to have brought the Bubonic Plague and its even more deadly variant the Pneumonic Plague that killed about a quarter of the European population in the 14th century.

The result of all of this was that cities brought with them a far higher disease load. According to the medical historian Roy Porter [3], until recent times mortality was so bad that cities never replenished their own populations, needing instead a constant influx of potential new disease victims from the countryside to keep them running.

On the other hand, exposure to such warrens of disease bred some resistance, leading to a Darwinian winnowing out of those with no natural resistance. Dying young, they were not able to pass on this lack of resistance to their offspring, at least according to present theory.

World population increased from perhaps five million before agriculture was developed to about seven billion and counting today. This massive population change was only possible because of the extensive development and application of agriculture and this too has had its own impacts on health. Large scale irrigation is necessary to get the most from land and crops in order to feed burgeoning populations but this also brings with it diseases from parasites like leeches that carry diseases such as bilharzia and schistosomiasis.

Though the development of agriculture allowed larger populations to be sustained it could also produce a diet low in proteins and vitamins leading to stunted growth and greater susceptibility to diseases like scurvy. In the transition from nomadic to the more agricultural Neolithic culture, people for a while actually grew shorter in stature and in some ways became less healthy.

Clearance of forests for agriculture often left behind a multitude of little water holes in which mosquitoes could breed. Mosquitoes carry malaria; as agriculture spread north from Africa then so did malaria.

There is another major source of health problems that has arisen from the growth of agriculture. This has to do with the farming of animals that can also be a source of infections. Even today the proximity of chickens, pigs and people, particularly in rural China, allows new strains of flu to jump species. Humans now share over fifty microorganism-based diseases with dogs, cows, sheep, pigs, horses and chickens.

Tuberculosis and small pox came from cows, whilst pigs and ducks gave us continually varying versions of influenza. One of these was the Spanish Flu that has so far been the worst flu pandemic ever, killing over 60 million after the First World War. Horses gave us the common cold and dogs gave us measles.

The more developed mankind became, the more trade routes were opened. Travellers from foreign lands brought with them diseases to parts of the world where resistance had not been developed. Measles and small pox drove the Aztecs and Incas almost to extinction when the Spanish Conquistadors arrived. Though it is not certain, it may be that these American peoples returned the favour by bequeathing syphilis to the Spanish invaders who then took it back to Europe.

Whilst modern man may have subdued many of these diseases and, in the case of small pox, perhaps even eliminated it in the 'wild' if not in the laboratory, a major new factor has emerged which will allow the more rapid spread of disease. The ease and reach of air transport make the spread of new diseases, or new strains of old diseases, very rapid.

It is not just new infectious diseases which come to a species at the top of the food chain. Other more recent epidemics of ill health come from our affluence: heart disease, diabetes, lung cancer, emphysema, cirrhosis of the liver and so on.

Modern Medicine

With all these diseases that have come from man's success in mastering his environment, as well as the usual ages-old scourges, modern medicine has had a hard road to travel.

Medicine has been largely ineffective for the great part of its history but how good is it now? Let's start by looking at some of its definite successes.

Antiseptic surgery

One of the biggest medical advances was by Lister in 1865 who used carbolic acid, at the time a means of reducing smells from sewers, to sterilise wounds, stitches and instruments to kill germs. Germ theory was a new concept at the time. Germs themselves were thought to be the invisible causes of diseases rather than vapours and miasmas as previously thought.

Suspecting these microscopic agents could move through the air, and wearing gloves to prevent him transmitting them by touch to the patient, Lister nevertheless was in the practice of operating while wearing a blue frock coat made stiff with congealed blood from previous operations and dissections (he must have made a charming dinner guest). Perhaps we should be forgiving of this foible bearing in mind that his work ushered in safe surgery and thus saved many millions of lives. Coupled with this, his more scientific approach began to move medicine towards saving more patients than it killed for the first time in its long history.

Antibiotics

Modern antibiotics have saved perhaps hundreds of millions of lives though some treatments based on their properties have been around for several thousand years. Treatments arose from the observation that biological substances, usually based on mould or plant extracts, could combat disease. Nowadays it is believed these antibiotic agents combat the microorganisms that are responsible for bacterial based diseases like tonsillitis and wound infections.

The antibiotic that has saved the most lives is penicillin and there's a good chance you'll have heard of the man who is regarded as discovering its antibiotic properties: Alexander Fleming.

Fleming and Lister are justifiably lauded because of the many of millions of lives their work ultimately helped save.

[That's undoubtedly impressive, but if we're looking for the overall winner in the saving of lives stakes and not just in medicine then it has to be Norman Borlaug, of course.

Chances are, you've never heard of him even though he's credited as saving over a billion people from starvation and so is arguably mankind's greatest hero.

Borlaug, an American biologist and Nobel laureate, is regarded as the Father of the Green Revolution that drove the development and transfer of new Western agricultural techniques to the developing world. These included things like high yielding grains, hybridisation of seeds and much else.

Don't worry if you don't know about him as for some reason he rarely gets the press he deserves. A cynic might say this is because none of the lives he saved were in the developed world. Save a child in the US or Europe of Japan and you make the newspapers, save a few hundred million children somewhere else and it's: ...Norman who?]

Smoking

The discovery of the link between smoking and lung cancer by Richard Doll and Austin Bradford Hill in the early 1950s led to the conclusion that smoking reduced life expectancy by about ten years for smokers, though more recent studies suggest it's actually nearer fourteen years. It is estimated that had smoking been banned altogether in 1950 it would have added six years to overall mean life expectancy.

Though medicine is not actually curing patients of the diseases of smoking such as lung cancer, it is giving them the knowledge to prompt them to stop smoking or to never develop the habit in the first place. Smoking rates as a result have declined by about a half in some developed countries such as the US, though it continues to rise in the developing world.

Medical advice to stop smoking has, and will continue, to save millions of lives.

Interestingly, Hippocratic medicine was associated with two gods: Hygieia who was the goddess of cleanliness, hygiene and the rational way of life, and Asclepius who was associated with the physician healing of the sick.

This is an example of where modern physicians have managed to persuade Hygieia to do all the heavy lifting.

Small pox

This is undoubtedly one of the success stories of modern medicine, the technique of vaccination supposedly being invented by Jenner in 1796 using pus from the lesions of milkmaids suffering from the much milder cowpox. Exposure to cow pox in this way made people immune from small pox.

In fact, the Turks were using similar inoculation techniques in the 1700s, perhaps even earlier. As noted by Lady Mary Wortley Montagu, a great beauty who had herself been badly disfigured by smallpox, Turkish practitioners put the scabs of smallpox sufferers into nutshells. They used these nutshells to scratch the skin of those to be inoculated, thus smearing infected fluid into the new cut. Lady Montagu managed to persuade the Princess of Wales to have her two young daughters inoculated in this way in 1722, setting off a vogue for inoculation where previously there had been considerable resistance.

After a long running concerted global vaccination programme, smallpox was declared eradicated in 1980. As we will see in Chapter 14 on the risks of science, the use of the word 'eradicated' is actually shooting rather high.

Vaccination for smallpox is undoubtedly a medical success story. For example, deaths from smallpox in Sweden dropped a thousand-fold. As smallpox was so prevalent in the 18th and 19th century, it's thought that its treatment may have been responsible for half the increase in life expectancy up to 1850.

So, medicine has had a number of clear successes in terms of saving hundreds of millions of lives directly or indirectly but sometimes it gets vastly more credit in terms of life saving than it is due. We'll look at one major and little appreciated example of this now.

Though medicine has many very desirable uses, such as in alleviating pain and treating non-fatal illnesses, deferment of death is one of the most important criteria for measuring efficacy. How good has modern medicine been in those terms?

For thousands of years, until about 1865, doctors were fixated on techniques such as bloodletting. After all, the most prestigious medical journal in the UK, perhaps even the world, is still called 'The Lancet'. The immoderate use of this instrument, coupled with obliviousness of the need for sterility or even basic cleanliness, killed many more than it ever saved. Causing sick, weakened people to vomit and empty their colons, the other standbys of Hippocratic medicine, didn't really help much either.

However, as these old methods were discarded things seemed to get a lot better. Between 1900 and 2000 in the developed Western World life expectancy went from 45 to 75 years. Medicine for most of the twentieth century was happy for people to believe this was due to their better anaesthetic, sterilisation and surgical techniques and to new medicines.

In fact, the contribution of modern medicine to this increased life expectancy is only a fraction of this. One estimate of the effect on modern medicine on the increase of life expectancy is that it was responsible for at most 20% of the total increase [2] and this was mainly due to a reduction in mortality of babies after birth.

If it wasn't modern medicine then what were the most significant factors responsible for this increasing longevity?

One reason was restrictions on air pollution. The latter has always been a problem in coal burning urban populations. The Great Smog of London in December of 1952 is thought to have carried off over 12,000 people alone. The introduction of smokeless fuel and other alternatives to coal radically reduced premature deaths from non-smoking related respiratory conditions.

However, the main factor responsible for increased life expectancy is the implementation of better sewage systems that prevented drinking water being contaminated by human faeces.

Awareness of this arises from the work of Thomas McKeown back in 1976 [4]. McKeown was an epidemiologist who drew on a unique source of data that recorded what people supposedly died from. From 1837, and unique amongst the countries of the world, in England and Wales the cause of death was systematically recorded on the person's death certificate.

McKeown's work indicated for the first time that better sanitation was primarily responsible for longer life.

He showed that from 1850 to 1971 deaths due to water and food borne diseases were alone responsible for a reduction in death rates by 21%. This broke down as follows:

Cholera, dysentery, diarrhoea 10.8%
Tuberculosis (non-respiratory) 4.6%
Typhus, typhoid 6.0%

This improvement can indeed be explained away by better sewerage (improvements in water and sewerage were introduced in the 1860s) and food handling, but McKeown also found that 40% of the reduction in death rates was due to an unexplained decline in mortality from airborne diseases. These break down as follows:

TB (respiratory) 17.5%

Flu, pneumonia, bronchitis 9.9%
Whooping cough 2.6%
Measles 2.1%
Scarlet fever and diphtheria (these weren't differentiated until later) 6.2%
Smallpox 1.6%
Infections of ear, pharynx, larynx 2%

You may be thinking that the lion's share of this reduction was due to modern medicine and the vaccinations and treatments for whooping cough, diphtheria and measles, but in fact that is not the case.

Death rates per million from respiratory tuberculosis dropped by over 85% (from nearly 4000 deaths per million people per year) between 1840 and 1945, even though the streptomycin treatment for the disease was not introduced until 1947. The BCG vaccination against tuberculosis had been developed in 1921 but its widespread introduction had been delayed as there were, indeed still are, doubts about its efficacy.

Scarlet fever death rates in children reduced to about a twentieth of what they had been between 1865 and the introduction of the protonsil treatment in 1935.

Whooping cough decreased from over 1400 deaths per million children per year, to a fifth of this before inoculation was even introduced.

Death rates from measles (1200 deaths per million children per year) decreased by over 80% before the remedy, limited to treatment for the disease's secondary infections only, was started in 1935. The immunisation for this disease only became available in the 1950s but has had little or no effect on the death rates as they had already fallen below 50 per million from a high of about 1200 per million in the 1880s.

In Britain the decline in deaths from diphtheria did seem to come after antitoxin treatment was introduced but it is possible this is coincidence as in other countries where introduction of the vaccination was delayed, for example in some American States, the reduction in mortality did not coincide with the vaccination programme.

Something made these and other big airborne killers much less effective but it wasn't medicine and it also wasn't better sewerage.

Reducing the overall death rate by over 40% is a colossal deal and dwarfs any other medical intervention in human history, but why it happened is still a mystery. Why few people are aware of this incredibly beneficial if mysterious boon is perhaps an even greater mystery.

One explanation for this improvement, and one that McKeown himself favoured, was that perhaps people became naturally more resistant to these diseases through better diet. This view seems nowadays dismissed by epidemiologists as there is some evidence to the contrary. Wrigly and Scofield [5] showed that in pre-industrial Britain standards of living increased but mortality worsened, then vice versa. Only in the 19th century did they both move in the same direction.

McKeown believed there was very little evidence of reduced exposure to disease causing this drop. People were as much exposed to whooping cough and streptococcus (cause of scarlet fever) as ever yet didn't die from them in anywhere near the numbers they did.

Some believe that perhaps the diseases themselves somehow lost their lethality, scarlet fever might be an example of this, or some other unidentified environmental factor was responsible. It is easy to explain the reduction in waterborne diseases due to better sewerage in this way, but not airborne diseases. Indeed, and we're straying into the grounds of speculation here, if there is another unsuspected factor responsible for the reduction in lethality of airborne diseases, then maybe sewers weren't the sole or even the main reason for the reduction in deaths from waterborne diseases, as is the conventional wisdom today.

Another explanation for some of this improvement may be as simple as 'micro-sanitation'. In essence this means better hygiene: a gradual awareness of the need for individuals to wash their hands more frequently, the increased use of soap, more domestic cleaning, covering up of food, reducing flies and vermin, and even the use of nappies for longer periods following birth. Again, however, one wonders how these could have such a major effect on airborne infection.

A more recent theory is that bacteria have 'virulence genes' which dictate how infectious and deadly they are. Perhaps these somehow got modified by some unknown factor in the 1800s [8].

Another possibility is that increased resistance to these diseases came from 'the ever-varying state of the immunological constitution of the herd'. In other words, the weak ones died off, leaving only the disease resistant strong to pass this trait on to their offspring.

Earliest records of births and deaths (but not causes of death) can be found in Sweden from 1751 and France from 1800. Generally, these showed death rates had already been falling from the beginning of the 19th century and from perhaps even as early as 1750. Drawing on his own work and the fragmentary data from before, McKeown came to the conclusion that over 85% of the total reduction in death rates from the early 1700s was due to the decline either in the prevalence or lethality of infectious diseases.

Reviewing McKeown's work, Emily Grundy [7] says that further studies, though based on less hard documentary evidence than McKeown, have shown that mortality increased in the 17th century. McKeown's 19th century figures, grim though they are, reflect an improvement on this which then gets even better as we head into the 20th century. One wonders if this up and down is only half of a long-term cycle, with diseases in synchrony repeatedly getting worse and better over time for a reason we have not even come close to identifying or explaining.

Some say the improvements revealed by McKeown are multi-factorial with better diet and cleanliness and Darwinian winnowing out all combining to increase life expectancy by a jaw dropping 12 years for airborne diseases, 30 years for waterborne and airborne diseases combined.

McKeown himself thought that, as all the diseases declined in mortality in the same way, it would be 'incredible' if it was down to a simultaneous change in their characters. He nevertheless followed through with this uncomfortable thought: 'if infectious deaths decreased because of a fortuitous change in virulence, they could quite easily increase again for the same reason.'

It is of course not just these diseases which seem to come and go. Bubonic plague killed between 25 and 60% of the European population in the 14th century, depending on who you listen to. Plague still erupts today yet it kills handfuls of people rather than in the tens of millions.

Like the reductions in deaths from airborne infectious diseases, it is not clear why this reduction happened. Were you even aware this was such a mystery?

And none of these reductions had anything to do with modern medicine.

Perhaps you think that, mysteries or not, there is really nothing to worry about nowadays. You may think our modern drugs like antibiotics would protect us from the worst effects of these diseases even if they were to suddenly become more prevalent and ferocious in their effects.

Perhaps not. In the early eighties 10% of the strains of Staphyloccocus aureus (cause of some respiratory diseases and skin infections like boils) produced enzymes which destroyed penicillin. Nowadays almost all do [8]. Micro-organisms are also adapting to protect themselves from such treatment in other ways, such as modifying cell structure so antibiotics can't penetrate, or developing ways of getting rid of them.

Even worse, it has been observed that the gene believed responsible for penicillin resistance can be transferred to other bacterial strains and even other organisms entirely.

Another major factor for reducing the effectiveness of antibiotics is overuse by man. Antibiotic resistance has only been made worse by putting antibiotics in animal feed, treating viral infections like flu and the common cold with them even though they rarely have any benefit, and by some countries like in the USA and the Middle East allowing their citizens to buy antibiotics over the counter. All this exposure gives greater chances for resistant bacterial strains to develop. The non-resistant strains get wiped out by the antibiotics leaving less competition for the resistant strains in that biological niche. Staphylococcus and streptococcus, ironically the ones responsible for Fleming's discovery of penicillin, are now making a comeback in multi-drug resistant forms. Tuberculosis is making a similar return.

There are other problems with antibiotics. These work because of their actions on cells, for example by breaking down the cell's wall or by interrupting its reproductive processes. Antibiotics are ineffective against viruses as these do not have the cellular structures and processes of bacteria. Though antibiotics kill unwanted bacteria they can also destroy useful bacteria as well, including Escherichia coli in the gut which aids digestion. By killing off these useful bacteria they may leave room for more harmful bacteria to take up the spaces left behind.

Bacteria are also developing new ways of infecting people. One example is necrotising fasciitis, the 'flesh eating' bacteria that originally came from a previously mild form of streptococcus.

For about 25 years up to 1960, antibiotic drugs seemed to have bacterial based diseases on the run. But for the past forty years, and despite the development of literally thousands of novel semi-synthetic antibacterial drugs, the tide has been turning back due to increasing drug resistance. Vancomycin is becoming the antibiotic of last resort and already there are signs that some bacteria are becoming resistant to that.

Our Heretic may float an alternative and rather more disturbing idea: perhaps antibiotic resistance is a myth. Could it be that diseases are just getting more lethal as part of a long cycle as mentioned before; that we are presently on the 'up' part of the cycle and so mankind may be heading back to Dickensian levels of death and disease?

Either way it may not matter. If a sudden resurgence in disease lethality doesn't get us, drug resistant bacteria might.

The world may be reaching a peak of life expectancy for its occupants.

Before we leave the ground-breaking work of McKeown it is worth pointing out that his findings did not show that all life-threatening diseases were declining. On the contrary, while all the infectious diseases were killing less people, big increases were seen in others. Cancer deaths per million quadrupled between 1850 and 1971, whilst those from cardiovascular disease tripled. This is thought to reflect the fact that as people were no longer being seen off by the infectious diseases, people lived long enough to be affected by the diseases of later life [6].

So, we've seen medicine's real big successes and also an example of where credit for success was undeserved, but how effective is the rest of modern medicine?

Surgery can be very effective but most people never need it at all and what surgery they do need is increasingly for improving their quality of life rather than the saving of it. For example, hip and knee replacement operations for joints damaged by the extra forces caused by obesity or because people living longer are having wear out their joints.

The main weapons in the modern physician's armoury are drugs and this is where we meet some real problems and gain some sense of the limitations of modern medical theory when faced with the complexity of the human body and the diseases that affect it.

Drugs

Surely all these wonderful drugs we have nowadays have been a tremendous boon to mankind and have saved many lives?

Life-saving claims for medicines need careful examination. Drugs do certain things which are beneficial to the human body in disease, but they inevitably have other effects which can be deleterious or even fatal. This is largely to do with the complexity of the human body and the processes that make it function. If you have a bad heart and you take a pill to treat your condition, then the active molecules from that pill don't all wind up in the heart; after absorption by the gut they find themselves in the blood stream where they are transported to every organ and tissue in the body. The effects of drugs on the non-targeted organs can be complex and subtle but nevertheless profound and can produce a plethora of adverse reactions. Next time you take any prescription pills read the enclosed documentation and have a look at the long list of the unwanted side effects. My own blood pressure pill, a medical favourite for treating hypertension, lists fifty such effects ranging from common ones like a tickly cough, skin rash and cramps, to more serious but less common problems like vertigo, depression, swelling of the gut, sexual inability, nail problems, ringing in the ears and even breast enlargement in men.

To examine the complexity underlying all drug treatment let's look first at the most commonly used medication of all: aspirin.

About 40,000 tonnes of aspirin are consumed a year. It is used to reduce temperature, relieve aches and pains and also to reduce inflammation. It is claimed to save countless lives [9]. You may therefore be surprised to hear that aspirin has never undergone a proper clinical trial of its effectiveness, aside from its use in heart disease as we will see in a minute. This is because aspirin, in one form or another has been around for hundreds and maybe thousands of years so has never had to undergo any of the clinical trials that recently developed medicines have to. With clinical trials costing tens or even hundreds of millions of dollars, it probably never will.

Despite the drug never being properly tested, anyone who takes it can tell you it can help lower temperature and often does make people feel better more quickly, but that in itself is not necessarily an indication of its efficacy in fighting the disease that caused the symptoms in the first place. High temperature is one of the body's mechanisms for combating viral or bacterial infection because such organisms tend to be more temperature sensitive than the body itself. For this reason, it is conceivable that aspirin may have actually caused more deaths than it saved in the Great Flu epidemic which killed between 50 and 100 million people after the First World War. This infected about a third of the world's then 1.5 billion population.

We just don't know if it helped more than it hindered; there is no hard evidence either way.

But aspirin has other uses and one of these illustrates the conservatism of medicine. Remember this is the profession that held onto bloodletting as a cure for just about everything for 2000 years because of adherence to Hippocratic theory.

Aspirin prevents clotting of the blood and this property is thought to be why it may be useful in preventing coronary heart disease, despite one of its side effects being to promote haemorrhaging, particularly in the stomach. As with any drug, the whole body is treated even though one is often only aiming for a single organ or system.

A doctor in the Midwest of America in the 1950s called Laurence Craven had been calling for the use of aspirin as a prophylactic (preventative) to help avoid coronary heart disease developing and had even started his own trial where he persuaded 8000 of his patients to take part. Heroic though his efforts were, the study was flawed as he did not use a control group and so did not take into account either the placebo effect or of patient non-compliance (patients may tell their doctor they are taking the medicine but often they just don't).

Generally, his work was ignored, not because of his shaky methodology but because there was no theory explaining how it might work. Medicine has the same hang-up as the hard sciences in believing theory first, however threadbare that may be, before experiment or observation. Indeed, Hippocratic medicine had a solid two-thousand-year history of doing just that.

Such professional resistance delayed the use of aspirin in this way for decades and may have cost millions of lives. When theory caught up, in other words it provided an explanation for the effect in its own terms, large clinical trials finally began.

These appeared to show that in patients who had already had a heart attack, aspirin reduced their chances of dying in the next year by 25%. This effect is a little less dramatic than at first sight because 88% of people survive these subsequent heart attacks even without taking the drug. Saving a quarter of the 12% who die means that aspirin only saves 3%. Nevertheless, as heart attacks are so prevalent this could, if used properly, save 100,000 lives a year worldwide. In addition, when given to a large group of patients with angina, it decreased the chance of a heart attack by 50%.

So, in this one condition aspirin has been properly assessed but it hasn't in any of the many others it is used to treat.

It's not just aspirin that hasn't been properly assessed in this way. Most of the drugs prescribed today came into use before clinical trials were really developed. This means that most of the drugs that doctors use, have not been tested for their effectiveness against the diseases they are prescribed for. This limitation also applies to many of the surgical techniques that are used in operating theatres around the world.

A study published in 1963 in Medical Care looking at family doctors across Britain showed that only 9% of their prescriptions had had their effectiveness or otherwise experimentally established. The situation is better now as more randomised controlled trials (RCT) have been undertaken of accepted therapies. Some more recent studies suggest treatments backed up by such trials may now be as high as 49% [9], though there are considerable variations from speciality to speciality. A 1998 study on paediatric surgery indicated only 26% of operational interventions were RCT based. In ophthalmic surgery it was 34%, obstetrics and gynaecology 42%.

Perhaps you are trying to take comfort in the thought that even though these drugs or procedures have not been properly tested, that surely doctors wouldn't prescribe them if they found from their own experience that they didn't actually work. I must however point out again the profession's two-thousand-year long adherence to bloodletting. One should never underestimate the siren call of theory even when set against one's own hands-on experience. If you don't get the results that theory and its august adherents say you should, then it takes a strong personality not to fall into the trap of assuming it's you that's doing it wrong.

Let's look at another drug to illustrate another point about the role of theory in medicine. As well as causing the medical profession to reject drugs that work but are not supported by theory, theory also leads them to too quickly adopt drugs that actually don't work because the theory behind them is wrong. An example of this is the anti-arrythmia drugs used to prevent irregular heartbeats which could lead to death, at least according to theory prevalent in the 1970s. By 1979, 12 million prescriptions a year were being written for these drugs in the US alone.

It wasn't until a trial years later in 1987 that severe problems were revealed with the anti-arrythmia drugs flecainide, encainide and moricizine. This was a controlled trial where half the patients were given one of the drugs, whilst half were given a sugar pill placebo. The trial nearly failed right at the start because doctors were so convinced that these drugs worked that they felt it unethical and indeed cruel to give half the patients in the studies placebos instead.

Luckily these objections were overcome but the trial still had to be prematurely terminated, though this time for the entirely opposite reason. It had rapidly become evident that the drug, while stilling those errant heartbeats as theory mandated, was killing more patients than it helped. In fact, two of the drugs in the trial are estimated to have killed 50,000 people in America alone [9]. The third drug was also found in a later trial to kill more than it cured.

Dispiritingly, cardiologists ignored these results and continued to prescribe these drugs for years or switched to other versions of the drugs made by other companies but not included in the trial. As Thomas Moore in his book 'Deadly Medicines' asked: 'How much evidence was enough to persuade doctors to abandon a theory that had been accepted without proof in the first place?"

A more contemporary example of the profession being misled by theory is the use of Hormone Replacement Therapy (HRT) in post-menopausal women. Surely, the theory went, replacing hormones in older women with those found in younger ones would be good? When, finally, a full randomised control trial was performed it again had to be halted early because the higher mortality from breast cancer in HRT treated patients became evident. In the UK alone, it has been estimated that HRT induced 2000 extra cases of breast cancer per year.

Let's take a break for a second from what may appear to be sustained criticism about medics' and medical scientists' slavish adherence to whatever the prevailing theory was at the time. Other scientists like some physicists may adhere to theory because of its mathematical elegance or other trifling reasons (described in earlier chapters) but medics generally adhere to theory out of desperation. Theirs is a difficult and at times heroic profession because of the terrible complexities they face every day of their working lives. As we will soon see, patients have widely different responses to the drugs they are given, many of which are barely effective in the first place and can produce unwanted consequences in terms of other forms of ill health (this is called iatrogenic disease which is defined as disease 'brought forth by the healer'). Iatrogenesis may account for as much as one third of the patients in hospitals [10,11]

On top of all that, diseases may come in multiples, each of which can assume a wide spectrum of severity. Drugs which may make one disease better may make another worse. When multiple drug therapies are used there is also the possibility of ill effects arising from the interactions between the drugs.

Medics therefore often have to make life or death or otherwise life changing decisions based on poor quality information often using not terribly good therapies, the actual usefulness of which may never even have been established in the first place.

In such a sea of turmoil a nice bit of supposedly hard-edged theory can be like a life belt. And what else of solidity has a hard-pressed medic to cling onto, even if the certainty it offers is illusory?

We've described how medicines that have been around for a long time might not actually be useful, but surely modern medicines with their extensive clinical trials must be better? Unfortunately, the truth is that new drugs, jumping though they do through all sorts of hoops to get regulatory approval, may still get onto the market doing little or no good. Chapter and verse on this subject can be found in the excellent book by physician and science writer Ben Goldacre [12] but here's a summary of some ways in which ineffective drugs are still getting through the regulatory process with depressing regularity.

We need to say something about clinical trials first, though. Randomised control trials are great when it comes to horrible diseases like cancer because you usually have 'hard' endpoints: the patient dies or they don't. RCTs are more difficult to set up if the effect they are looking for is difficult to objectively measure. 'Feeling better' is a much softer endpoint, for example, in determining whether selective serotonin re-uptake inhibitor (SSRI) based drugs are effective in the treatment of depression. Diseases not having hard endpoints makes assessing drug efficacy very difficult.

Nevertheless, initial trials of new experimental drugs may produce positive results (the drug works) or negative results (it doesn't). Surely that must help the regulators and the medical profession work out whether the drug should made more widely available or not?

Unfortunately not. Trials where the results are negative have only a quarter of the chance of being accepted for publication in the medical and scientific journals as opposed to results showing a positive result.

This selective publication effect artificially, and sometimes disastrously, weights perceptions of drug effectiveness.

In 2008 a retrospective examination was made of 74 studies of new antidepressant drugs introduced between 1987 and 2004. 38 were positive, 36 were negative. 37 of the 38 positive trials were published, but only 3 out of 36 of the negative trials made it into print. Full knowledge of the results might lead one to the opinion that the new antidepressants were generally pretty useless (over and above the placebo effect), but this selective publication effect would suggest a very high degree of efficacy [13].

Getting scientific work published is a highly competitive business; as indeed is the scientific publishing market itself. Journals publishing the most interesting papers that are subsequently cited by other researchers in their own papers get the most prestige and the most subscriptions. Negative results are far less sexy than positive results. The mantra spouted by lecturers to students of any form of science that: 'Negative results are as important as positive ones', is in real life quite naive.

This selective process is not limited to medicine. The non-appearance of negative results means that we often don't get to know when many theories fail in science generally. We suffer a kind of selective amnesia where only theories that produce positive (supportive) results remain while many scientific papers showing contrary findings drop by the wayside.

Even worse, and another reason why rubbish drugs manage to get on the market, is that many of the positive results that get into press may not be repeatable. In other words, other researchers repeating the same study don't get the same results. Begley and Ellis [14] tried to replicate 53 laboratory studies looking at possible cancer drugs and failed in 47 cases. The reasons for this may be many, though Begley and Ellis, perhaps charitably, thought it was mainly due to fluke findings being more sexy than negative ones. What this means is that if you are using statistics to show an effect, one typically sets a significance level such that the positive results happening by chance would do so only 5% of the time. That means one in twenty studies, by chance alone, will show an effect even if there isn't one. The predisposition of journals to publish positive results means that these flukish ones are more likely to get published than the more plentiful and representative negative ones.

It is also possible that this selective publication bias does not just lie with the editors of scientific and medical journals exclusively. Researchers themselves may just not bother submitting negative findings because they may be worried that they got the methods wrong (they see everyone else apparently producing positive results) or they themselves perceive the results to be too boring to publish (again, because they see so many exciting positive findings being published).

The result of all this is that it is mainly the statistical flukes that are making it through the publication process.

Another big factor that helps bring ineffective drugs to market is that most trials are performed by the drug companies themselves. Bearing in mind that hundreds of millions of dollars are required to bring a drug to market then it is unsurprising that some element of bias, conscious or otherwise, can creep in.

The regulatory process itself is also deeply flawed. Regulators obtain results of trials from early in the regulatory process but often not from the later ones. Companies can conduct as many studies as they want but are under no obligation to publish ones that produce negative results. A recent study [15] on five major drug types, such as anti-depressants, looked at over 500 trials and found that industry funded studies found positive effects in 85% of cases. Where the research was publicly funded (and therefore less likely to be affected by any commercial concerns) this figure dropped to 50%.

Another study found a twenty-fold difference between commercial and public funded studies on the efficacy of statins in the lowering of cholesterol and the subsequent reduction of heart attacks [16].

Even highly adverse side effects can be hidden, voluntarily or involuntarily, in this way.

Another major factor in getting ineffective drugs into our bodies is that, incredibly, in order to get regulatory approval drug companies need only show their drugs are better than placebo. They don't have to show they are better than other existing, and perhaps more effective, drugs. New drugs can be less efficacious and have more side effects than existing drugs but as long as they're better than placebo, they can get regulatory approval.

Old drugs superseded by better drugs which have been shown to be more effective with fewer side effects can also quite legally continue to be marketed.

One final mechanism by which inappropriate drugs may find their way into patients is worth mentioning though it is not quite so unequivocally bad as the above. Once a drug is licensed, for example for treating prostate cancer, individual doctors can nevertheless prescribe it for any other conditions that it has not been properly tested for. This is called going 'off-license'. Sometimes this is understandable because getting a regulatory license for use in another disease is expensive and takes a lot of time so a company may never do it. Usually this is because the disease is too rare for them to make a profit from the sales once the cost of regulation has been taken into account. It may actually be good for the other disease but, without hard evidence, off-license prescribing is a matter of guess work on the part of the doctor.

The reason some doctors are willing to do this can be seen with children who generally make up a small percentage of patients. Drugs may be tested on an adult form of the disease only because there is a much larger potential market, but it may not make economic sense to perform another study for its use in children as would be required by the regulatory bodies. Prescribing doctors can nevertheless make that leap of faith and prescribe it in kids. The doctor may be right or they may be wrong. Nobody knows or can know until large numbers of unbiased clinical trials are done but, in the vast majority of cases, these will never be performed.

Desperate diseases call for desperate measures and one's heart can only go out to a medic faced with a very sick child and the only possible effective treatment is a drug tested in adults.

Tough job indeed!

Why is developing a drug so expensive? There are many factors which mount up to the typical cost of hundreds of millions of dollars. Here are some of the reasons:

Countless tests are required to make sure that, when given in pill form, the active compound gets into the blood of the patient. You then have to test to make sure it doesn't get excreted from the body too quickly before it can have an effect.

You also have to try and ascertain that it doesn't cause cancer, though this can be problematic as it takes cancer years to develop. Instead so-called surrogate tests are used like the Ames test to see if a new drug causes mutations in bacteria in the test tube.

You need to try to ascertain if the drug is toxic to other organs and body systems as a drug in the bloodstream will go everywhere and not just to the disease site you want to target. This will require extensive animal testing. Not only do you need to see if the molecule is toxic in its original form but you also have to establish if its breakdown products can be harmful.

For regulatory purposes you also need to establish the LD50, the dose level at which 50% of the subjects the drug is given to die. That can be ascertained on animals, not humans, thankfully.

One big limitation of any animal testing of drugs is that it often doesn't tell you what will happen in humans. According to one source [17] only 10% of drugs that are both tested on and actually seem to work in animals also work in humans. The reverse is also true. Animals and humans respond in different ways to different substances. For example, mice have a 100,000-fold greater resistance to some common inflammatory toxins than humans. On the other hand, even things like chocolate can be toxic to some animals but not to humans.

As an aside, it's worth mentioning that there is some selectivity as to which drugs pharmaceutical companies will try to develop. Considerable preference is given to drugs designed for long term administration because those regimes are the most lucrative for the companies. A company can still make money if, as in the case of antibiotics, the drug has to be given for a few weeks but they're never going to do all that research and pay all that money to develop a medicine that is used only once. A single one-off pill to cure a disease is never going to appear, even if such a thing were possible in the first place.

If drugs are to be administered long term then animal testing has to be done for much longer and this also adds to the expense.

Then, after the animal work, they do the first trials (Phase I) in humans to begin to establish dose ranges, starting with a handful of healthy volunteers and using very low doses that are then increased. Sometimes in the past a drug was administered simultaneously to all the volunteers in the trial but there is less appetite for that following the disastrous administration of the potentially immuno-therapeutic drug TGN 1412 to half a dozen healthy subjects simultaneously. They all rapidly succumbed to massive and multiple organ failure. Some hints of this possibility had been found in a previous laboratory study but this had not reached print.

Phase II studies are generally performed on a few hundred patients who actually suffer from the targeted disease. Like Phase I, which is trying to establish a safe dose range, this is looking for biological effects. Good or bad. In Phase III trials larger numbers are required and thousands of patients may be recruited but this time with some alternative treatment or placebo being used for comparison purposes in half the patients.

Determinations of how effective the drug actually is can now be made, though it must be pointed out that very often the disease itself, particularly if it is chronic in nature, may not be the outcome measured. Instead a surrogate often has to be used. For example, high cholesterol is associated with heart disease and statin drugs are designed to lower the levels of cholesterol. It is the lowering of the surrogate marker, cholesterol, which is usually the efficacy measure, not whether the patient is less likely to develop heart disease because it would take many years to establish the latter.

Using a single surrogate marker of disease in this way is quite limited. The causes of heart attacks are multi-factorial (high blood pressure, familial history, age, ethnicity, diabetes, obesity, and goodness knows what other factors we are oblivious of at the moment) and it doesn't mean that lowering cholesterol will actually lower your chances of a heart attack. In fact, as we saw with the anti-arrhythmia drug, lowering the surrogate marker of arrhythmias actually increased the chances of heart attack, killing perhaps 100,000 worldwide in the process. Targeting single components of complex systems that we don't understand can sometimes make things much worse.

Another example of this involves the drug torcetrapid made by Pfizer which was designed to affect cholesterol levels [18]. Cholesterol isn't simply bad for you. In fact, cholesterol is a chemical vital for life and comes in different forms. Some, from the viewpoint of our present limited state of knowledge, are labelled as 'good' (HDL cholesterol), some as 'bad' (LDL cholesterol).

Scientists have been studying the metabolism of cholesterol and its variants for a hundred years since Nikolai Anichkov first proposed a link between it and the build-up of fatty deposits (atheromatous plaques) in the arteries. Ultimately these plaques can lead to blockage of the artery and produce angina, or they rupture and can cause death.

Cholesterol metabolism is very complex and it takes a score of stages involving intermediate biochemical compounds to produce the final versions of cholesterol.

The new drug had been shown in a small sample clinical trial to reduce 'bad' LDL whilst increasing 'good' HDL. With such good initial results, the drug company went on a full-scale Phase III clinical trial. What could possibly go wrong?

One may not be surprised by now to find out that things didn't go as planned. Rather than preventing cardiovascular disease it produced a 60% increase in overall mortality as well as more chest pain and heart failure. The trial was therefore prematurely terminated. It is an example of why something like 40% of trials fail at the Phase III trial level. Pfizer had spent $1 billion and involved 25,000 subjects in total in the trials.

The failure of the drug, and the theory that produced it, is another typical example of the failures of the reductionist approach to science (as described in Chapter 7). Reductionism, the way of modern science generally, holds that complex systems, such as you and me, can be broken down into their individual parts. These can be individually analysed and, in the case of 'bad' cholesterol, can be targeted with a drug. In the USA alone, $100 billion a year is spent on drug development using just such a reductionist approach.

In this case of torcetrapid, an intricate cholesterol metabolism was broken down into all its complex stages, with each stage being carefully investigated. Usually all this sort of work is done in the massively simplified environment of the test tube.

The problem is that cholesterol metabolism, complex though it may be, isn't the only process in the body. There are countless others and it is the unexpected interaction of cholesterol metabolism with these which is thought to be the cause of the poor outcome with torcetrapid. In this case the unexpected consequences were significant effects on blood pressure.

The body functions via many such bio-chemical pathways and can be overwhelmingly complex with one pathway effecting a second, and then the first and second together effecting a third, and so on. On top of this many, perhaps all, of these pathways vary to some degree from individual to individual. This is why patients' responses to drugs can be so variable and why many drugs only work in a fraction of those to whom they are administered.

Sometimes we strike lucky, though. Pfizer has another cholesterol lowering drug called Lipitor, a statin, that is prescribed more than any other branded drug in the USA. It is now believed that rather than working via its primary effect of reducing cholesterol in the blood, it is actually the secondary effect of reducing inflammation that is the effective mechanism.

Even in this successful drug, vast complexity lurks just under the surface. This is manifested by the range of unwanted side effects it can produce, ranging from headache which is found in over 10% of patients, to less prevalent problems (in less than 10% of patients) including among others: abdominal pain, constipation, nausea, chest pain, urinary infection, rash, arthritis, sinusitis, bronchitis.

And other common substances interact adversely with this drug. For example, one serious side effect prevalent in less than 1% of patients is myopathy which is a muscle disease causing weakness. Niacin (vitamin B3) in combination with the drug increases the risk of this. Grapefruit juice can also increase overdose toxicity, meaning it can increase the potency of the drug, as indeed it can with many other drugs including some blood pressure medications.

With all the above in mind it is a miracle that any drugs are useful at all and we should not be surprised that a magic bullet to cure disease doesn't exist. We should also not be surprised that, if we include the costs of the other drugs that a given company tries to develop but that finally fail, the costs of bringing a successful drug to market is now well over $3 billion.

The complexity we have discussed above also means that even if a drug is successful its actual effects will be highly variable from one individual to another and may ultimately not be generally efficacious at all. Though this is difficult to say for sure, at least one source has reckoned that around 85% of drugs approved by European regulators provide little or no benefits [19].

Astra-Zeneca and GlaxoSmithKline have indicated they are reducing their research funding to develop drugs for illnesses involving the brain because it is just too complicated with too many pathways and processes involved [19].

Indeed, the whole business of the drug treatment of disease is getting so complex one wonders if most of the best medicines have already been found. Many of the drugs we use today came from folk medicines where treatments were based on observation and experience, though this too could often produce ineffective or harmful therapies.

Many of our modern medicines that are effective have generally not been found from theory but from observation of often unexpected effects. Drugs developed for one disease are also often found fortuitously to be good for the treatment of some other disease, if not for the target disease. Theory is then quickly cobbled together to explain why they work in that way, though this may be just as fallacious as the original theory. Viagra, for example, was originally developed to treat chest pain and its effect on erectile dysfunction was a chance observation. It was also observed to be beneficial for the effects of altitude sickness. These beneficial effects were unanticipated at the time but are now buttressed by theory produced after the event. More examples of unexpected drug benefits are to follow in the next section.

Unexpected beneficial effects, and underwhelming expected effects, are a common feature of drug development. Indeed, animal and human trials of active molecules are often fishing expeditions looking for such fortuitous therapeutic bounties.

Let's look at how medicinal treatment has evolved to cope with disease to illustrate some of the above points. There are quite a few diseases out there and life, even without disease, is short so let's confine ourselves to the consideration of the most feared disease of all.

Cancer

Cancer comes from the Greek word for crab to symbolise its crab-like projections into surrounding healthy tissue. It was first described by the Greek physician Galen over 1800 years ago. One third of us will die from it or its secondary consequences. Its complexity, both structural and metabolic, is why it is so difficult to deal with.

Its causes also appear to be manifold. Some of these have been suspected for centuries, though these have waited until the 20th century for theories to be developed to explain why [8]. Even as early as 1700 Bernardino Ramazzini had noticed the much higher prevalence of breast cancer in nuns compared to women who had suckled many infants. Similarly, more nasal cancer in snuff takers was observed by John Hill in 1761 (snuff is a tobacco product), scrotal cancer in young chimney sweeps by Percivall Pott in 1775. The latter came about because children cleaning chimneys were often nude because of the filth and their later washing was insufficient to remove the chemical-containing soot from their scrotal crevices. Despite Pott's observation, the practice continued for nearly a hundred years.

In a similar vein, Joseph Bell (Arthur Conan Doyle's old teacher whose deductive methods were the model for those employed by Doyle's fictional character Sherlock Holmes) described 'paraffin cancer' of the skin in 1876.

Roughly speaking, and according to theory, 30-35% of cancers are tobacco related; not just of the lung but also of the bladder, kidney, mouth, larynx and pancreas too. 40% of cancers may be food related, for example due to the chemicals in preserved meats like bacon; to high fat, low fibre, fried and grilled foods and to nitrate and nitrite rich diets (BBQ anyone?); to low vitamin C diets and to high alcohol consumption.

The hundred-fold greater incidence of liver cancer in some parts of Africa compared to the UK is thought to be from the presence of poisonous products of fungus in groundnuts which are part of the staple diet, though an association with the increased exposure to Hepatitis B may also be a factor.

Increased exposure to apple brandy (Calvados) is thought to be the cause of increased oesophageal cancer in Normandy. Higher rates of stomach cancers in Japan are thought to be due to the consumption of salted fish and pickled foods. On the other hand, the reduced incidence of prostate cancer in Japan is thought to be due to the low amounts of red meat in the diet compared to other countries in the developed world.

When people of one ethnicity move to a foreign country then the incidence of such disease tends to become similar to the predominant ethnic group in the new country, tending to support this diet related theory.

Other sources of cancer may be ionising radiation related. Some are viral related, for example the human papilloma virus and cervical cancer. Both are almost unknown in nuns because of lack of sexual exposure to the virus. Swings and roundabouts though, bearing in mind their higher incidence of breast cancer.

Some people also seem to have a higher genetic predisposition to certain types of cancer.

Whatever the cause of the cancer, the current theory is that it arises from mutations in the duplication of the DNA and hence the subsequent construction of abnormal proteins based on the 'recipe' the DNA provides. Our DNA suffers perhaps 10,000 damaging events per cell per day. Some may be caused by ionising radiation from cosmic rays or other background radiation or by the action of naturally occurring atoms or molecules or ions called free radicals. Red wine and fresh vegetables, the present theory goes, can intercept and destroy these free radicals, hence the current vogue for the 'Mediterranean Diet'.

Some of these mutations don't seem to matter but, if they do, then they usually kill off the cell before it can propagate and pass the mutation on. However, sometimes these mutations can survive and continue to produce debased proteins and that can be the start of cancer.

More complexity is introduced because there are also mechanisms that are thought to control the growth and differentiation of cells (proto-oncogenes), or code for enzymes that control transcription and repair of DNA (tumour suppressor genes). Disruption of these can also result in cancer.

Cancer cells that grow but do not spread are considered benign, though they can still kill. If the cancer spreads to other tissue then it is said to be malignant, for examples cancers that spread to the bone following breast or prostate cancer if not caught speedily enough.

Huge effort is devoted to finding new compounds to treat cancer. Latterly, many compounds have been developed by first hypothesising their mode of action based on some current theory, but in fact the most effective cancer drugs have not been found this way. Chance and luck in conjunction with careful observation are keys, as with so much of medicine that is actually effective.

One of the first effective cancer drugs was found by chance by a physicist called Barnett Rosenberg who noticed similarities in the patterns of cell division and the patterns of iron filings sprinkled around the poles of a bar magnet. He decided to study the effect of magnetism and electricity on cells and fortuitously used platinum electrodes. They were thought, wrongly, to be inert and so would not have a chemical effect on the experiment. Rosenberg noticed that cell division was unexpectedly prevented. What had happened was that light had fortuitously changed the nature of one of the platinum compounds and it was this that was found to be effective in inhibiting cell division.

Though the results were found by chance it is to Rosenberg's credit that he immediately saw the potential of the technique in treating cancer. He and Loretta van Camp produced an active compound called cisplatin (trade name Neoplatin with a subsequent analogue called carboplatin) and tested it on solid tumours in mice to great effect. Tumours shrank.

Cisplatin and its analogues have revolutionised the treatment of teratoma (testicular cancer) with two-year survival times rising from 20% of treated cases to ultimately 90-95%.

The first useful cure for leukaemia came originally from using folic acid to try to treat the condition. Folic acid had previously been found to have success against bacteria. It was hoped folic acid might be useful against many types of cancer but in fact pure folic acid wasn't. Luckily, a batch of test drug supposedly containing it actually contained a chemical derivative. It was this folic acid residue, called Teropterin, which caused tumour regression by disrupting DNA synthesis, at least according to a retrospectively constructed theory.

It was found that, surprisingly, pure folic acid actually seemed to speed up the disease process and hasten deaths of leukaemia sufferers. The theory that folic acid would be good to treat the disease was quickly taken through a 180-degree turn, the emphasis for developing new drugs now being on inhibiting the production of folic acid and its derivatives. Various drugs were tried without significant success until one called aminopterin was produced. Although patients lived longer with this drug only about 1% made a full recovery.

But there were many side effects, including compromise of the immune system. Further folic acid derivatives were investigated to find ones that produced less serious side effects. These were then coupled with other chemotherapeutic drugs in various combinations until, by trial and error, combinations were found which produced much greater efficacy. We are now at the point where, after treatment, patients with childhood acute lymphoblastic leukaemia have an 80% chance of complete remission after four years.

So, theory has played little or no part in the development of our most effective anti-cancer drugs but there is one significant exception: 5-fluorouracil made by the chemist Charles Heidelberger in 1957. Uracil was a compound taken up more readily by cancer cells than normal tissue. Heidelberger believed uracil was a precursor of thymine which is a major component in DNA. He believed that, just as changing a hydrogen atom in acetic acid produced the highly poisonous fluoroacetic acid, that doing the same thing to uracil might have a similar effect and the cancerous cell would be killed. And indeed, that has proved to be the case and the drug is now used particularly with cancers of the colon and rectum.

Incidentally, as with many other therapies, there are many examples of drugs being developed to treat cancer based on theory but which are subsequently found to be ineffective. However, they are subsequently found to be very useful in treating other diseases. One example of this is azidothymidine (AZT) which was unsuccessfully developed to treat cancer but was later found to be useful in treating HIV infections.

Thus, theory in medicine often takes the back seat to chance discovery as has been the case throughout history.

It was the twenty-metre-long scroll, known as the Ebers papyrus, which showed that as early as 1500 BC the Egyptians had about 800 herbal remedies. The Chinese in about 200 BC had 365 drugs. Some were effective, some weren't.

Using chemical analysis and animal and clinical trials to work out what were the active chemicals in effective herbal medicines took until the 20th Century. The subsequent search for new biologically active constituents of plants and micro-organisms (often obtained from soil) using large scale screening programmes didn't start until the 1940s.

This non-theory based approach almost immediately yielded many potent anti-bacterial and anti-tumour agents.

The haphazard but ultimately fruitful nature of such research can be seen in the case of the drug called vinblastine. This came from a plant called the Madagascar periwinkle which was held, by observation based ancient medicine, to be useful in diabetes. To everyone's surprise, testing the extract on animals killed them. This was because they were more prone to infections due to a reduction in their white blood cells. It was then realised that vinblastine could be very useful in treating leukaemia.

The yew tree was used by ancient Greeks and Romans as a poison and so was another rather interesting lead for modern medical research. This yielded the anti-cancer agent taxol, something particularly effective in advanced ovarian cancer.

Investigations of soil sample extracts also yielded streptomyces parvullus in 1953, helping to produce a drug found to be effective in a specific type of kidney tumour. Mitomycin C is another soil sample derived chemical found to be useful in treating large solid tumours.

Samples from the sea have also produced many drugs [8]. For example, in the mid-1960s, George Petit showed that perhaps as many as 10% of marine organisms produced chemicals that inhibited the growth of mouse leukaemia. Drugs called Bryostatins came from the sea snail Dolabella auricularia and initial trials on these suggest they may have significant anti-tumour properties.

To date only a tiny fraction of organisms has been properly investigated but the future potential for producing active agents is huge.

Of course, as well as there being naturally occurring compounds that may have medical benefits, there are also many compounds which may cause ill effects. Bruce Ames who invented the Ames test for mutagenicity said: 'There are more natural carcinogens by weight in a cup of coffee than potentially carcinogenic synthetic pesticide residues in the average US diet in a year, and there are still a thousand known chemicals in roasted coffee that have not been tested'[8].

We'll end this chapter on medicine by looking at one recently discovered but effective treatment for an old disease. It is an admirable piece of work by a brave man who overturned orthodoxy and so faced the inevitable vilification in the process.

It's a nice story because, for once, there is a happy ending for the heretic.

Gastric and duodenal ulcers used to cause great suffering with perhaps 10% of the population suffering from them at some point in their lives and with a quarter of a million dying from them each year worldwide.

Universal medical wisdom had it that these were caused by things like stress and spicy food. Treatment for stomach ulcer had been by palliative medicines (in other words they eased not cured) like Tagamet and Zantac, that reduced the production of gastric acid. These drugs were vast sources of income for the drug companies that made them. However, these drugs hardly ever seemed to be effective for long and so many sufferers ended up being treated by surgery. As with all surgery this sometimes killed, usually because of excessive bleeding or infection or complications arising from using an anaesthetic.

Diagnosis of the condition was made by endoscopy. This needed a tube to be fed down into the stomach and a sample of the stomach lining taken. So many of these endoscopies were performed that it became a vast industry. Even better for the purposes of income generation, the treatments used hardly ever cured the problem and it usually recurred every two years. Each time another expensive endoscopy was required.

In 1982 an Australian gastro-enterologist called Barry Marshall discovered the bacteria Heliobacter pylori (or Campobacter pylori as it was known then) in the stomachs of people suffering from peptic ulcers and severe gastritis [20].

Marshall figured that if ulcers were in fact caused by bacteria then antibacterial agents could eliminate these very simply, thus putting an end to all the surgery and the therapies that had treated the symptoms rather than the disease itself. Marshall used antibacterial agents on his own patients with great success.

That's when his troubles began. He submitted his work to an Australian medical conference and had it rejected. His submission was rated in the bottom three of nearly 70 papers. When he did finally present his work on the international stage he was met with derision. This derision was to last for many years. The prevailing medical theory was that bacteria could not survive in the acid of the stomach so of course that meant his theory must be nonsense.

Many of his colleagues called Marshall a quack.

In frustration, and in one of the more admirable traditions of medicine, he tried his theory out on himself. Though not suffering from an ulcer, he drank a hell's brew of the bacteria that he believed caused it and, sure enough, became rather ill. Ten days later an endoscopy showed gastritis and the presence of the bacteria.

Using his own antibiotic treatment, he was then able to cure himself.

Nevertheless, it still took twelve more years until the US National Institutes of Health published an opinion that the great majority of recurrent ulcers were due to the bacteria and could be treated with antibiotics. Millions of chronic sufferers had their misery suddenly and permanently relieved. Potentially fatal surgical procedures were avoided as surgery for uncomplicated ulcers became obsolete.

70,000 less people a year globally die of the disease compared to before

1990, and even more would be saved if they could afford the antibiotics to cure it.

The happy ending? In 2005 Marshall won a Nobel Prize in the category 'Physiology of Medicine'.

Summary

This chapter has tried to show that medicine has had a very chequered history. For thousands of years and no matter how absurd some of the concepts and theories employed by bygone medics, these were believed in implicitly despite many of the medics' own experiences and observations to the contrary.

And modern medicine, though far more rigorous in terms of seeking hard evidence, has done exactly the same thing over the last hundred years. The beliefs and theories may be different but are held with similar levels of trust and confidence.

Is today's medics' faith in modern theory as misplaced as their predecessors in Hippocratic theory? Modern theory really springs from the 17th century ideas of the body as nuts and bolts and the scientific way forward was seen as being that of reductionism. The latter, dividing the body into its constituent parts and carefully investigating the functions of these in as close to isolation as possible, has been the hallmark of modern medical science.

But, of course, the complexity and myriad interactions between different organs and systems in the body that are never in real life isolated is always going to set limits on the effectiveness of such theory in combating disease.

The most effective advances in treatment or prevention have come from observation, whether it be from watching blood-soaked surgeons running higher mortality maternity wards, or observing that milkmaids rarely caught smallpox, or simply noticing that smokers died much younger than non-smokers. Theory to explain this that was also acceptable to the scientific view then current often did not come until many years later.

Surgery can be very effective but this arises from refining techniques based on actual hands-on experience compounded over generations of surgeons. Theory takes second place to knowledge gained from such experience.

Many of the effective modern-day drugs are based on folk medicines that were found to be effective by trial and error and experience. Of the rest, most of the effective ones were developed for some other disease that theory indicated they would be efficacious for. In the event they usually weren't but were fortuitously found to be effective in other conditions.

Theory has played little part in the development of the most effective drugs and indeed, in some cases, has even stymied their development. This raises the question as to what present treatments are dictated by completely erroneous theories? What is the present-day equivalent of bleeding and leeching dictated by belief in the four humours?

Incremental improvements in treatment of disease are going on all the time but again come from experience and observation rather than being guided by theory. In this way drug doses and drug combinations can be refined and optimised so that over time outcome is improved.

And because of the complexity of the body and its interactions, every therapy has unintended consequences best illustrated by the range of each drug's side effects. This complexity also degrades the uniformity of response and is the reason why no drug works beneficially in all patients with a given condition.

The rate of progress in medicine when it comes to developing effective drugs to treat disease seems to be slowing, at least when guided by theory. Such new drugs nowadays generally make only small improvements or no improvements at all on what has come before. On the other hand, and very much not driven by theory, the search for naturally occurring 'active' molecules that can help fight disease gives up hope that effective medicines are still out there to be discovered.

So, diseases of the body are a huge and complex challenge for modern medicine but there's something out there that's arguably even more challenging to the profession. We'll look at that next.

11

Medicine and the Mind

The aim of psychoanalysis is to relieve people of their neurotic unhappiness so that they can be normally unhappy.

Sigmund Freud (supposedly)

And what science had revealed was this: Prior to treatment, patients diagnosed with schizophrenia, depression, and other psychiatric disorders do not suffer from any known "chemical imbalance". However, once a person is put on a psychiatric medication, which, in one manner or another, throws a wrench into the usual mechanics of a neuronal pathway, his or her brain begins to function, as Hyman observed, abnormally.

Robert Whitaker [1]

The travails of medicine when it comes to treating bodily illness may be seen as a consequence of the vast underlying complexity inherent in the body. Something similar is the case with psychiatry, medicine of the mind, because we have very little understanding of how the 'nuts and bolts' of the brain actually produces this thing we call the mind in the first place.

In this chapter we will look very briefly at the history of the treatment of mental illness and how ineffective such treatment has been. We'll show how the psychiatric profession is deeply divided into two camps. One side is trying to shoehorn the wide spectrum of the symptoms of mental illness into possibly quite fictitious diagnostic pigeonholes generated by committee and, at the stroke of a pen, turning tens of millions of us into sufferers overnight. The other side meanwhile isn't even sure that mental illnesses are entities themselves and instead are the product of external social factors rather than actual diseases that somehow originate internally.

Psychiatry, in short, is, and always has been, in a state of chronic turmoil.

Something that is not appreciated about psychiatry is just how much it owes to leprosy. This is because psychiatry commandeered leprosy's physical

196

infrastructure, and even perhaps its social stigma, when in around 1400 AD leprosy pretty much vanished from the Western world.

While the disease was still rampant, lepers were kept hidden in places away from, or on the edges of, towns and communities. These 'leprosaria' or 'cities of the damned' numbered about 19,000, with 43 in one diocese of Paris alone. In the UK in the 12th century, and for a population of then only 1.5 million, there were 220 leper houses [2].

Incidentally, the explanation for leprosy's disappearance from northern shores is not known. Perhaps it bears some similarity to the other infectious diseases mentioned in the previous chapter that also radically declined for reasons not yet satisfactorily explained.

As far as leprosy is concerned there is, however, a simple and reasonably believable explanation. Perhaps the policy of segregating lepers from the healthy population actually worked.

Perhaps also the end of the Crusades very much reduced the passage of the bacterium to the West from the Middle and Far East. That sounds credible but doesn't quite explain why leprosy did not get worse again with the growth of international trade from the 16th century onward and before the development of drug treatments in the 1940s.

So that's one mystery. How it's transmitted from human to human is another.

In any event, the decline in leprosy meant the empty leprosaria were just the right places to put the insane instead. That may sound rough but bear in mind that in 1400 AD the insane were often publicly whipped, so confining them in the vacated cities of the damned might be considered a big step forward.

Incidentally, though the public whipping may have become less frequent in the following centuries, there were still many horrors to which the mentally ill were subjected. Medics were still in thrall to Hippocrates and his humours and indeed, even until 1783, the treatment mantra in the Bethlem Royal Hospital in London (and source of the world 'Bedlam') was: bleed them, make them vomit and purge them.

It's possible these therapies didn't work on the insane at all but they kept being applied for thousands of years anyway.

Historically there have been essentially three main trends in psychiatric thought:

That madness could be explained in physical terms. In other words, it had an organic cause

That it had a psychological source and so the problem came from the thought processes themselves

That it was either inexplicable, or was due to interference by God or demons, and the only cures were by magic

The Egyptians were exponents of the first as can be seen from their views on hysteria, a word derived from the Greek hysteron, meaning uterus. Convinced this was a uniquely female condition, they believed that hysteria was caused by the uterus moving around the woman's body. The obvious solution, to them at least, was to fumigate the vagina with smoke to lure the uterus back to its proper position.

The Greeks themselves advocated marriage and sexual intercourse as a way of putting a stop to the wandering of the miscreant uterus.

The Greeks also made some of the first recorded attempts at psychotherapy. These involved dream states, perhaps using drugs to induce them. The historical record suggests they were the first to consider that the brain was the site of mental functions [3].

Later, the Greeks and Romans thought the mind was actually sited in the diaphragm. One of these believers was the unfortunately named Soranus who, as well as doing some original work in gynaecology, used fairly gentle and restful methods to treat psychiatric patients. This stood in marked contrast to one of his students called Celsus who tried to frighten them back to health with loud noises.

Hippocrates, again with his humours, thought that excessive cold, heat or damp could cause madness. He also thought that intelligence came from the inspiration of air. Depression he put down to an accumulation of black bile.

He and his follows also made some of the first attempts to classify mental illness into the categories of epilepsy, mania, melancholia and paranoia. To the Romans, the latter term meant mental deterioration and not the meaning of unwarranted distrust in others that it has today.

The Roman Aretaus in about 100AD was the first to carefully observe the mentally ill and even went so far as to do follow up studies on them which was something that had never been thought of before. This revealed that mania and depression could occur in the same patient.

The Hebrews subscribed to the view that madness was an infliction from God via evil spirits and indeed the Bible describes examples of depression (in Saul), catatonic excitement (a form of extreme agitation) and epilepsy. Epilepsy is now considered a condition that can be caused by both anatomic but more often physiological abnormalities, but it was considered a psychiatric condition well into the 19th Century. Nebuchadnezzar's lycanthropy, the delusion that he was a wolf, is also described in the Bible. Indeed, as early as AD 490 Jerusalem had a hospital for the mentally ill.

St. Augustine in about AD 400 showed, from his own personal experience, how introspection and self-analysis were vital for psychological knowledge; for example, he used it to help him understand his grief when a close friend died.

The subsequent Dark Ages brought no new developments in the field, or at least none that were recorded. The Renaissance saw a blossoming of both knowledge and more liberal viewpoints but not when it came to mental illness. Instead witches were generally blamed for the condition. In 1487 the Malleus Maleficarum (The Witches Hammer) was issued and became the textbook for the Inquisition and its treatment of witches. It made many interesting stipulations. For example, when brought before the Inquisition a woman accused of witchcraft had to have her pubic hair shaved off lest the devil hide amongst it.

Though directed at witches, the book took in many of the mentally ill as well; as a consequence, hundreds of thousands of women and children were burnt at the stake.

It took until 1563 for the Malleus to be rebutted by the brave Johann Weyer. Brave because the book had been directed not only at witches but at heretics as well. In his book De Praestigiis Daemonum (The Deception of Demons) Weyer says of witch burnings: 'Almost all the theologians are silent regarding this godlessness. Doctors tolerate it...those illnesses whose origins are attributed to witches come from natural causes'.

In the 17th century, better observational approaches were being taken to mental illness by Thomas Hobbes and William Harvey, famous for his early studies on the heart. Thomas Sydenham described the clinical manifestations of hysteria and how they could simulate almost all forms of organic disease. He also recognised that men as well as women could suffer from it, though for men he preferred to use the term hypochondria.

It also took until the 17th Century for Spinoza to put forward the notion that psychological phenomena could be as important as organic processes in producing mental illness.

So even over three hundred years ago the battle lines between the nature of mental illness being organic or inorganic were being drawn for a battle that is still inconclusive to this day.

Meanwhile doctors where still bleeding and purging away, supplemented now by dousing the insane in ice water. Also, the novel technique of skull blistering was used in the case of King George III who may have been suffering from a depressive psychosis and/or porphyria, an enzyme disorder that can also produce mental disturbances.

In the UK you'd have to go private to get that sort of treatment nowadays.

However, by the end of the 18th Century some investigators were looking for the sites of mental illness in the brain at autopsy. Peoples' brains are highly variable; the adult human brain can easily vary in weight form 800g to 1600g, and there is also considerable variation in local anatomy from person to person. This means that if someone was different in terms of their behaviour when they were alive then it's easy to find something different about their brain in death, though the two may not be in any way related.

Phrenology was also popular at this time. Adherents of this believed that very specific brain areas had specific functions such a 'love of offspring', 'wit' and 'veneration'. These different areas had different sizes and the health or otherwise of these could be inferred from feeling the lumps and bumps on the head.

Despite all this, the brain was not regarded as the sole cause of psychological symptoms, particularly as far as women were concerned what with their wayward genitalia and reproductive systems. William Cullen, inventor of the word 'neurosis', considered this condition a result of an excess of sexuality that had not found fruition in childbirth. Young widows, he felt, were particularly prone to this condition. As with so much of this early autopsy work where individual variation and small sample sizes often facilitated finding what people were looking for, his dissection studies led him to believe that: 'Observations of the dead bodies of patients labouring under hysterics, shew that in most of them the ovaria are affected. These are liable to a turgescence, which gives an irritability to the system, and hence a want of venereal pleasure is assigned as a very common cause of the disease.'

In other words: if you're neurotic due to swollen ovaries, then what you really need is sex.

Then again, that's advice that would make just about anyone feel better.

Though forms of classification and, albeit erroneous, diagnosis were slowly being developed, the prognosis for patients with mental illness was bleak. Most were considered incurable, purging and bleeding notwithstanding. However, in 1803 Johan Reil produced the first systematic study of psychotherapies, many of which still followed the old Greek duality of either shutting patients in quiet, darkened rooms or exposing them to sudden loud noises. Reil proposed new therapies based on drama, music and occupational activities and was against the common use of opiates in such patients as he felt they could induce more mental illness than they cured.

Boldly for this precious 19th Century man whose profound belief in the purity of women drove him to impotence, he also prescribed prostitutes for other men with that condition; not so much for the prostitutes to use their professional wiles to cure it, as to show the patients that few women lived up to the romantic ideal of womanhood.

By the end of the century there had been advances in understanding the basics of brain chemistry and in neuropsychology. The latter is somewhat like phrenology, in that it attempts to identify areas of the brain associated with specific functions, but uses better techniques than feeling someone's head

As we saw in Chapter 7, the First World War brought advances in neuropsychology with so many apparently fit and able young men receiving bullet and shrapnel injuries to the head. Psychologists were able to correlate site of injury, in terms of where the bullet entered, with physical or psychological deficits. Though better than nothing, the results were nevertheless highly prone to error. High velocity bullets don't generally take out one little bit of the brain; the shock wave from the impact affects all of it to some degree. Coupled with this, the large-scale carnage of the war meant that turnover of troops was so great that recruiters weren't choosy, so many recruits had mental and physical disabilities even before the flying metal hit their head.

All this meant that correlating sites of damaged brain with psychological effects was, ironically, a rather hit or miss affair.

The 1917 flu epidemic also showed that infectious agents could cause neurological and psychological damage, again suggesting organic causes for some mental illnesses at least. Vitamin B deficiency was identified as causing psychotic symptoms; metabolic disorders such as galactosemia and phenylketonuria were found to be responsible for mental retardation. Some of these disorders were also shown to be hereditary.

However, in Vienna in 1908 an entirely new way of looking at mental illness was being developed. Sigmund Freud was a neurologist who decided to focus on, as he put it: ''functional' nervous diseases with a view to overcoming the impotence which had so far characterised their medical treatment'.

Freud didn't invent the idea of the 'unconscious', those parts of the mind that supposedly produce ideas and perceptions and needs which may not be directly within the individual's conscious control. However, his theories were liberating as they involved the idea that one's own sexual desires, perverse though they may appear to others (and don't they always), were a product of one's own unconscious and not some purposely acquired evil. This was quite contrary to the prevalent Christian view of the time.

Indeed, Freud believed that sexual repression, whether through society or religion, as well as sheer ignorance of sex, produced mental illness. By allowing the patient to talk freely he hoped to tease out the desires of the unconscious mind so the patient could become aware how these conflicted with the prohibitions of society. Once aware of these conflicts the patient might, if not be cured, at least be relieved of the symptoms this unperceived war had been creating; troubles often laid down by conflicts of sexuality in childhood that had subsequently shaped the personality.

Freud's new psychoanalytic approach became integrated into medicine. By the middle of the century the view of such disorders being essentially internal responses to external factors, rather than being organic in nature, became prevalent. This view was reinforced by the effects of the First World War where returning soldiers showed evidence that fear and stress themselves could produce apparent physical as well as mental ailments.

So popular did Freud's way of looking at things become that the number of psychiatrists in the United States increased six-fold in the thirty years after the war.

Freud's psychoanalysis was aimed at those with severe neurotic conditions to make their symptoms bearable but he did not have great faith in effecting actual cures. Certainly, he appears never to have anticipated the widespread application of his methods to essentially normal but unhappy people, particularly in the United States. Indeed, the rich in the US began to use psychiatrists as the equivalent of spiritual advisers, or as surrogate parents for their unhappy children and even, in some cases, for themselves.

As well as the 'talking cures' of psychoanalysis, new types of therapy were being tried. It was found that using insulin to send schizophrenic patients into coma for thirty hours or so (insulin shock) seemed to have beneficial effects for the ones who survived. However too many could not be brought back to consciousness so another form of 'shock' was sought, this one based on electrical shock as developed by Ugo Cerletti in 1938. His idea came from the since unsubstantiated observation that epileptics didn't develop schizophrenia. This led him to believe that inducing an epileptic-like fit might help cure the disease.

Cerletti experimented on hogs to ascertain the safe levels of voltage; in other words, levels that subdued them without killing them. He supplemented the shock treatment with curare-like drugs to reduce the induced muscular contractions that otherwise were so severe they could cause bone fractures.

Experiments showed that in fact electro-convulsive therapy (ECT or 'jump starting loonies' as I once heard a rather louche practitioner describe it) was actually found to be much more effective in treating severe depression though, for the effects to last, it needed to be supplemented with psychotherapy.

You would think that because ECT uses electricity it must be a relatively modern technique but that's probably not correct. The first ever use may have been by Scribonius Largus in AD 50 who treated the headaches of a Roman Emperor by using an electric eel.

Amnesia was, and indeed still is, a common result of ECT and this can last from weeks to months. Indeed, this led to one of the many theories as to why ECT or any other form of 'shock' treatment might work. Some thought that the electricity-induced amnesia blocked out the memory of the event that had precipitated the episode of depression. Others thought the shock may have satisfied some need for punishment on the part of the patient, or that the muscular contractions or the shock itself made the body think there was a threat to life and so it mobilised its defences which then cured the underlying problem. Another theory was that the electricity produced a negative feed-back to the brain's electrical system, so damping it down.

One other theory was that the patient was so fearful of the treatment that they simply 'escaped to health'.

Nobody knows why passing electricity through the brain seems to work in some instances. Cynics might wonder whether witchcraft and magic still dwell in the heart of psychiatry, though by other names.

ECT may sound bad enough but even more radical therapies were appearing. Moniz, a Portuguese neurologist, in 1935 used frontal lobotomy in cases where psychoses were resistant to ECT. This usually involved cutting out large sections of the frontal lobes of the brain and was the rage for about twenty years, despite the fact that the mortality rate from the therapy was 1-2% and it often left the patients acting like 'placid zombies'. Moniz won the Nobel Prize for this work in 1949, though by the 1970s his techniques were banned in many countries. At least 40,000 patients were lobotomised in the States alone. It was used to try to cure depression, schizophrenia and homosexuality. This latter 'condition', as we will see, was deeply troubling to psychiatry throughout the 20th century.

It is not well known that a battle has been raging within psychiatry for more than a century. It is around the issue of whether mental illnesses really are entities themselves, presumably arising from some sort of organic de-rangement, or whether they are essentially non-organic in origin and instead are a consequence of external factors causing a derangement in the way a person thinks and so behaves.

Convinced of the former proposition the 'categorisers' believe the broad spectrum of mental illness can be broken down into hundreds of discrete categories of disease. For the last thirty or so years the categorisers have been in the ascendant.

On the other side are the psychoanalysts from the Freudian tradition. These are sometimes labelled 'anti-psychiatrists' by the categorisers. The psychoanalytical view is perhaps exemplified by comments of the psychiatrist Karl Menninger, that there was 'only one class of mental illness- namely mental illness' [4]. The vogue for categorisation was 'restrictive and obstructive'.

He and others such as Ervin Goffman, Thomas Szasz, Thomas J. Scheff and R.D. Laing believed mental illness was a consequence of social problems and oppressive and stultifying family environments. For that reason, they campaigned against what they regarded as the causes of mental illness, namely social ills like poverty and racism.

Corroborating their beliefs was the uncomfortable fact, for categorisers at least, that there seemed to be no anatomical basis for mental illness in that the brains of sufferers were indistinguishable from those considered normal. Indeed, this was why 'anti-psychiatrists' argued that mental illnesses were not diseases at all; that a patient's 'craziness' as perceived by the psychiatrist was in fact only a measure of the social distance between the two. In other words, if the psychiatrist was in the patient's shoes they would see exactly where they were coming from.

Szasz, a qualified psychiatrist and psychoanalyst in his book 'The Myth of Mental Illness' essentially equated psychiatrists with astrologers and alchemists [5]. He felt that therapy should in fact take the form of a systematised study of human relationships rather than be regarded as a medical speciality at all.

Scheff argued that there was 'no rigorous knowledge of the cause, cure or even symptoms of functional mental disorders. Such knowledge as there is is clinical and intuitive, and thus not subject to verification by scientific methods' [6].

Even when it came to the severe psychotic condition schizophrenia, R.D. Laing regarded it neither as a condition or an illness but rather as a label and that it was a 'social adaption to a dysfunctional society'[7].

This approach became very popular in the 1950s and 60s as did the belief that mental illness could at least be alleviated by changing the patient's environment or how they dealt with it. That putting them in mental institutions solved nothing as the therapies offered, whether it be drugs or lobotomies or ECT, were generally ineffective. This is one of the main reasons why in the US between 1955 and 1994 the number of patients in state mental hospitals declined from 550,000 to only 80,000 [6].

In fact, this way of thinking dates back to the founding of American psychiatry in the 1800s, though in those days rather than being called psychiatrists, practitioners were called alienists. Alienists, however, believed that mental illness could be cured by taking patients out of their adverse environments. Sometimes these alienists would take the patients into their own homes and indeed this is ironically how mental asylums first came into being, though later they assumed a much more custodial function [6]. The more custodial they became the more harm they did, according to the alienists.

As well as seeing their income falling as fewer sufferers were being put into asylums in the second half of the 20th Century, things got even worse for the categorising psychiatrists who were the ones who generally ran these asylums. This followed from a study that had been undertaken by a radical Stanford psychologist called David R. Rosenhan who had become very concerned at how variable psychiatric diagnoses were. Often three psychiatrists examining one patient would produce three different diagnoses.

Rosenhan decided to mount a study in which normal subjects were admitted to a dozen mental hospitals. As far as ward staff were concerned the 'patient' had referred themselves because they had 'heard voices'. However, as soon as they arrived at the hospital these pseudo-patients stopped reporting experiencing these. From then on, they acted just as themselves. These pseudo-patients were kept in hospital in some cases for over fifty days and yet none were rumbled as being anything other than schizophrenics or, in one case, manic depressive. Their fellow patients were the only ones who thought that something funny was going on because they noticed the pseudo-patients taking so many notes.

Rosehan then went on to compound his heresy by arranging a study where hospital staff were informed that more of these pseudo-patients would be referred to their hospitals in the coming months. Forewarned, hospital staff managed to identify many of their patients as Rosehan's plants.

But, of course, Rosenhan had been lying; he had not referred any further pseudo-patients at all.

So, in Rosehan's first study psychiatrists had diagnosed people as being ill who weren't, in the second they had diagnosed people who were really suffering from mental illness as being normal.

This all supported Rosenhan's view that: 'any diagnostic process that lends itself so readily to massive errors of this sort cannot be a very reliable one'.

Smarting from assaults like this and seeing their business crumble and also being appalled by the anti-psychiatrists' suggestions that psychiatry wasn't even a science at all, there was a backlash by the categorising psychiatrists. And, to be fair, there were aspects of psychoanalysis which didn't cut the mustard either. That approach too had little or no success when it came to dealing with psychotic (usually meaning loss of contact with reality) conditions such as schizophrenia or bi-polar disorder.

Even with depressive illnesses where one might hope psychoanalysis would be effective there was no evidence that this was the case. Indeed, psychoanalysis has as its foundation the uniqueness of the patient and their problems. This means that the efficacy of any therapy psychiatrists applied could never really be established as there would be no identical, or even equivalent, control patients available to compare non-treatment with.

Without such corrective feedback showing whether their therapies worked or not, quackery developed and this saw the reputation of psychiatry plummet. There's evidence of this in the Hollywood films of the 60s and 70s where psychoanalysts with their nebulous 'talking cures' are almost universally portrayed as figures for mockery.

To try to make the profession look more scientific, an attempt was made to definitively categorise all the mental illnesses from which mankind suffered, taking as read that these represented real disease entities, to a very large part at least.

Like psychoanalysis, the categorisation approach has been around for about a hundred years, the very early attempts of the Romans notwithstanding. Emil Kraepelin had been one of the first to try to systematically categorise mental illness back in the late 1800s. He proposed dividing the psychotic disorders of unknown origin into dementia praecox (now called schizophrenia), manic-depressive (now called bi-polar) and paranoia. Even at the time many criticised this categorisation as being too simplistic but these categories remain to this day.

The Heretic would argue that right there was modern psychiatry's first big mistake; that these categorisations divide a spectrum of disorder into three arbitrary categories. Over time these categorisations became hard and fast and psychiatrists who use them begin to think that schizophrenia is itself a completely separate disease from bi-polar disorder, whereas that may not be the case at all.

Kraepelin's approach soon lost ground when Freud appeared at the beginning of the 20th century because his psychoanalytic approach at least offered the possibility of an amelioration of symptoms. The categorisation approach offered nothing in the way of treatment.

Nevertheless, more concerted attempts were made to formalise diagnostic categories in the 1950's with the first edition of the Diagnostic and Statistical Manual of Mental Disorders (DSM-I). This was succeeded by DSM-II in the 1960's. Both were essentially academic works which had little impact on the clinical psychiatry. DSM-III was a different matter and its impact on the clinical practice of psychiatry cannot be underestimated, as we will see later.

Problems inherent with the modern categorisation approach were there from the start. For example, DSM-I had homosexuality identified as a mental illness. A cynic might suggest that, in the face of losing so many patients from their mental hospitals, the profession was trying at the stroke of a pen to make as many people mentally ill as possible.

After a terrible furore by the Gay Liberation Front and other organisations, the second edition of the DSM was altered to remove it but only after the psychiatric profession had taken a referendum amongst themselves to decide the issue. Instead homosexuality was only to be considered a mental illness if the 'sufferer' was troubled by their sexual orientation. In fact, an entirely new term had to be coined for this phenomenon: homodysphilia.

That the profession had to vote amongst themselves to decide whether something was a mental illness or not did nothing to improve the status of psychiatry as a science, or as anything else for that matter.

At this point it's worth pointing out an emerging pattern: that of the cyclical nature of the belief systems in psychiatry. The alienists, forerunners of the anti-psychiatrists in the mid-1800s, lost ground to Kraepelin who in turn lost ground to Freud who then lost ground again to the categorisers. Over the space of just over 100 years the pendulum of psychiatric orthodoxy has swung back and forth, then back and forth again.

Shifting sands.

But in the 1970s the categorisation approach was about to begin its triumphant rise, though not without at times vitriolic disagreements amongst the profession.

The most famous latter-day categorist, and successor to Kraepelin, was Robert Spitzer who was made head of the DSM III Task Force. In this endeavour 25 committees were set up to more closely categorise mental illnesses.

As mentioned, one major impetus for the DSM was the poor reproducibility of diagnoses when the same patient was examined independently by two or more psychiatrists. Few studies had actually been performed to measure this but those that had been were far from encouraging. In one study after the Second World War, three psychiatrists were found to agree on a diagnosis for a given patient only twenty percent of the time. Even when only two psychiatrists were involved, they agreed less than half the time. This led the author of the study, Phillip Ash, to the conclusion: 'The extremity and frequency of disagreement make it quite legitimate to raise the question of the reliability of the diagnostic system' [9].

Spitzer in DSM III (which was finally published in 1980) set out to improve the situation by 'the development of specific inclusion and exclusion criteria for each diagnosis that the clinician is required to use, regardless of his own personal concept of the disorder'. Psychoanalysts being the way they were, this was an ambition not dissimilar to herding cats.

In DSM III the ambition had been to firm the decisions regarding the categories by using hard data from properly controlled studies. Unfortunately, this failed when it was realised there were no such studies and no such data.

Instead the twenty-five committees were set up to try to make these categorisations by discussion and, hopefully, consensus. And indeed, there was much discussion over the five years of the project with much heat being generated in the process. Whilst clear categorisation was indeed the final product in the published DSM-III, this belies the massive difficulties these committees had in reaching consensus. Without hard study data, all they had to rely on was anecdote and argument.

As a result of this process, and despite massive criticism from many of the psychiatric profession at the time, DSM III resulted in whole new disorders suddenly emerging, such as attention deficit hyperactivity disorder (ADHD): a childhood syndrome. This was a big change as DSM-I hadn't even had separate categories for any childhood disorders.

There was even suddenly a 'Tobacco Use Disorder' later renamed 'Tobacco Dependence' in response to the furore and ridicule caused by tens of millions of 'normal' people being suddenly medicalised overnight.

The conditions of autism, post-traumatic stress disorder (PTSD), anorexia nervosa, bulimia and panic disorders also suddenly appeared and grew 'empires of sufferers' [10].

'Neurosis', a mainstay of psychoanalysis was dropped as a separate condition. Howard Berk, a senior member and post holder of the American Psychiatric Association, wrote on this and other matters: 'the elimination of the past [he was referring to longstanding psychoanalytic practice] by the DSM-III Task Force can be compared to the director of a national museum destroying his Rembrandts, Goyas, Utrillos, van Goghs, etc, because he believes his collection of Comic Strip type Warhol's (or what have you) is of greater relevance' [6].

'Neurotic depression' became no longer neurotic and was no longer considered a depression. 'Hysteria', beloved by medics and psychiatrists for thousands of years, became factitious (in other words artificial or a scam) and equally applicable to both sexes.

Theodore Blau, President of the American Psychological Association said that the Board of his professional body objected 'strongly' to these new definitions of mental disorders and pointed out that of seventeen major diagnostic classes at least ten had no known cause or origin; in other words, perhaps they weren't separate entities at all. The proposed definition of mental disorder was 'arbitrary'. Later Blau said: 'It would be irresponsible to produce a diagnostic system without an empirical scientific data base'.

Handy checklists of symptoms were nevertheless provided by the DSM, a certain number of which the patient had to display to be officially diagnosed as having the disease. 202 specific mental conditions were described in DSM III whilst DSM II had managed only 180.

The issue of personality disorders also caused conflict amongst the profession. These are supposedly responsible for patterns of behaviour and ways of thinking that are very different from those prevailing or even mandated by society. In fact, if you're a heretic and are rather loud about your views then this is the category you might end up in.

Personality disorder is an amorphous and contentious concept that many psychiatrists had no time for but was nevertheless somehow given eleven different sub-types in DSM II, for example: Inadequate, Explosive, Hysterical, Anti-social. Spitzer, via DSM III, was keen to take out four of these like Inadequate and Explosive but felt it necessary to add five more: Schizotypal, Avoidant, Borderline, Dependent and Narcissistic.

Aaron T. Beck, a world leading expert on depression wrote that, in terms of personality disorder, this was a construct so artificial and removed from observables: 'That it is probably of little utility and, even worse, it is probably a misleading fiction' [6].

And this is all for a disorder for which there was, and still is, little in the way of hard experimental evidence even for its existence at all. Nevertheless, the categorisation has been persistent. Even today, in the draft version of DSM-5, Personality Disorder as a categorisation is still considered just as valid as other mental disorders.

Meanwhile homodysphilia wasn't cutting it anymore. Remember DSM II had ceased to stop regarding homosexuals as medically ill, unless they were unhappy about it, in which case they were. Even so, in a sense this was still separating out homosexuals. After all, heterosexuals who had problems enjoying sexual relations or concerns about their sexual identity weren't considered to suffer from a disorder and didn't have an impressive new word like homodysphilia coined to describe it. Why should homosexuals be so lucky?

So contentious did this issue become that one of the original members of the DSM-III's Psychosexual Disorders Advisory Committee, Richard Green, resigned. People pointed out that perhaps homosexuals who felt distress at their condition didn't do so because of some underlying disease but because society, certainly in the 1970s, hounded and ridiculed and discriminated against them. In fact, this rather chimed in with the old Freudian psychonalytical view where mental ill health was often seen as the result of external rather than internal factors. But this, of course, was anathema to the new categorisers like Spitzer.

Despite acrimonious debate, Spitzer's view prevailed and the condition still appeared in DSM III though, instead of being called homodysphilia, it was called 'ego-dystonic homosexuality'.

Whew, glad that's all sorted out then!

So, homosexuality was still at least indirectly pathologised (made to become a disease) because there was no corresponding ego-dystonic heterosexuality.

Thankfully even the category ego-dystonic homosexuality was removed in a later revision, namely DSM-III-R.

Interestingly, Spitzer later went on to claim that homosexuality could be fixed via 'reparative therapy' though later he retracted the paper he had published on the subject and even appeared in a video offering his profound apologies to the gay community.

This issue of the DSM's variable treatment of homosexuality illustrates a telling point. In this case at least, the identification or otherwise of a so-called mental disorder had been a function of time. What this means is that the categorisation of the 'illness' had changed as society's attitudes to homosexuality had changed. Who is to say that this isn't also true to some extent at least with other 'mental disorders' and that they are not set-in-stone absolute disease entities that the categorisers would have us believe, but rather that they change with time as societies attitudes towards them change?

It's not just a minority like homosexuals who have fallen foul of the categorisers' point of view. A slightly larger group, this time potentially comprising a little more than half of the entire human race, have similarly suffered. Women have also seen disorders originally ascribed solely to their gender change with time. DSM-III, in draft form and following on from the previous editions, was still banging on about the diagnosis of 'Gender Identity or Role Disorder of Childhood' which young girls disproportionately fell victim to in the eyes of the categorisers. Tom-boys, in other words, were pathologised.

The fact that girls might wish in some ways to be like boys may at that time have been quite easily explained without resorting to mental illness. Up until at least the time of DSM-III it was boys who had more power over their lives and were generally given more options in terms of how they lived it. That alone might explain why some little girls wanted to be like them.

The Women's movement at the time was growing strong and Spitzer was taken aback by the strength of their response to this draft categorisation. He and the Task Force were forced to respond and tomboyism was de-pathologised at the stroke of a pen. That sort of thing only became an actual disorder if the person, either a boy of a girl, persistently stated a desire to be of the opposite sex.

Again, these examples: hysteria, homosexuality and tomboyism, all illustrate the problem with categorisation of supposedly fixed and absolute diseases. If they really were like that then they wouldn't change with time and depend on the changing views of society. It also again supports the psycho-analysts' view that it is the external effects of society which produce much of what we regard as mental disorders rather than some innate disease process based on malfunctioning 'nuts and bolts' for which, incidentally, there is little or no physical evidence.

DSM-5 is the most recent edition of this work and is still only in draft form, but it too has provoked vitriolic debate. For example, it has (at the time of writing) bereavement lasting more than two weeks being classified as major depressive illness [11]. That and other changes serve to lower diagnostic thresholds, producing medicalisation of many people hitherto considered relatively untroubled by psychiatric illness. This, of course, suits the pharmaceutical companies down to the ground as they have more new diseases that they can sell people drugs to supposedly cure.

Even the British Psychological Society has recently described the most recent DSM's categorisations as being 'based largely on social norms and subjective judgement'.

There is a growing feeling nowadays that the DSM's simple disease model approach may be fallacious and that no specific disease entities underlie what are simply a range of symptoms; that categorisation of mental disease is itself a phantasm.

There are no other fields of science where truth has to be decided on by a vote. Had the results of these processes been unanimous, or even just clear, then maybe they would be legitimate. However, since the drafting of DSM-III everything has been the subject of great dispute and even drama. It has set one whole section of the psychiatric profession against another.

Yet somehow, despite this maelstrom, Spitzer and his team produced a 494-page manual which gave full definitions and diagnostic criteria for hundreds of supposedly discrete disease entities. DSM-I and DSM-II had been essentially academic works and had made little impact on psychiatrists or the practice of psychiatry but DSM-III was a game changer. The Manual (doesn't the very word smack of something concrete and real?) became a major cornerstone of modern clinical psychiatry worldwide and has sold millions of copies.

DSM-III, by channelling the diagnoses of the psychiatric profession into very discrete and perhaps even fictitious channels, has perhaps served to limit the way mental disorders are conceptualised.

Shoehorning patients into discrete categories for diagnostic or health insurance purposes (the DSM has had a vast effect on the health insurance industry) has become the norm but has left many psychiatrists struggling. Many resort to co-morbidities, in other words they diagnose their patients as having more than one of these specific categories of disease. Or instead they often use words such as 'atypical' to indicate the patient stubbornly resists being fitted into a nice, tight diagnostic category.

This limit was acknowledged by the co-chairs of DSM-5 who said: 'While diagnostic reliability has thrived, large scale epidemiological studies have underscored the inefficiency of DSM's criteria in accurately differentiating diagnostic syndromes.' [12]

Another problem was that DSM-III promoted the growth of checklist psychiatry. If patients ticked a certain number of boxes in one of the diagnostic checklists then the diagnosis was made. This relieved some psychiatrists of the need to spend time getting to know and trying to understand what might be troubling their patients.

Insurance companies love checklists. A psychiatrist taking a patient through a checklist does it far more quickly than if he or she were to spend time building a 'narrative history' of the patient as had been the practice for over a hundred years. This means the insurers have less to reimburse the psychiatrists (reimburse is a rather more delicate expression than the term pay which somehow makes this lofty process sound rather squalid).

So, there are big problems with the two main psychiatric approaches of psychoanalysis and categorisation. But what about drugs used to treat mental disturbances. Are they effective?

Indeed, one argument for the DSM approach is that its defined diagnostic criteria allow homogeneous patient groups to be identified so that proper double blinded, placebo-based drug trials can be mounted to determine the efficacy of any new drug.

Let's start with psychotic illnesses. Many antipsychotic drugs, known as neuroleptics or major tranquillisers, try to target a supposed excess in the dopamine system, a group of nerve cells in the brain. This system is theorised to be the cause of psychosis.

Perhaps surprisingly, given the supposedly discrete nature of disease categorisations, these drugs are also used in a very wide range of other mental illnesses including: autism, personality disorder, Tourette's syndrome, PTSD and obsessive-compulsive disorder. One begins to suspect that the mode of action is simply that of tranquillising the patient rather than having any effect on the disease itself, assuming such an entity exists at all. Indeed, few suggest these drugs actually cure anyone of anything; they just reduce some of the symptoms.

Many anti-psychotic drugs have been around for a long time and so have not had to undergo the current levels of regulatory testing to determine whether they really work or not. However, proper control studies are increasingly being performed. For example, anti-psychotics were commonly used in patients with intellectual disabilities who were prone to aggression in adulthood. One recent study found them to have no beneficial effect over placebo.

One reason these retrospective studies are being done now is to make sure they are actually useful because anti-psychotics have such severe adverse effects. Indeed, up to two thirds of patients stop taking them for that reason.

These side effects include dizziness, anxiety, Parkinsonism, dystonia (muscular contractions causing abnormal postures) and sexual dysfunction in both sexes. Some of these individual symptoms can occur in nearly half the patients taking some anti-psychotics.

That such drugs are treating individual symptoms is illustrated in bi-polar disorder where mood stabilisers like lithium may be effective in blunting the manic phase but most versions have no effect on the depressive phase. Similarly, anti-depressants may be used to try to blunt the depressive phase but they also have generally little or no positive effects on the manic phase. In fact, anti-depressants can themselves induce the mania phase.

How lithium blunts the manic phase is unknown.

Lithium, like many of our most effective medicines was fortuitously discovered, rather than its development being driven by theory. Indeed, it was originally taken to dissolve kidney stones and to treat gout. It was also, in the early days at least, an additive to the drink 7-UP. It was noticed as early as the 1880s that it also made people feel a sense of well-being and that it calmed some manic patients in asylums. This ultimately paved its way in psychiatric treatment for damping down the potentially harmful mania part of the bi-polar cycle.

This same type of chance discovery is true of another of the most effective drugs in psychiatry discovered in the 1950s. Antihistamines were developed as anti-allergy drugs but were found to have sedative side effects. This led to the development of a new anaesthetic that was then found to have anti-psychotic properties useful in quieting the manic phase of bipolar disorder. So effective was it found to be in schizophrenia that it ended ECT and lobotomies for the treatment of this condition. It has since been replaced by other drugs such as risperidone, whose mode of action is also unknown, but which causes fewer side effects such as 'tardive dyskinesia'. This manifested itself as Parkinson-like involuntary movements.

Anti-depressants are another example of a drug that is used to treat a wide range of disease categories, again suggesting their effects are on the symptoms rather than on any actual underlying disease. These other illnesses include obsessive compulsive disorder, migraines, ADHD, eating disorders and even snoring as well as much else.

The problem with trying to decide if treatments for psychiatric disease work is that there is a severe confound and that is the placebo effect. This is where even simply giving a patient something like a sugar pill, or even waving a magic wand over their heads, will make some patients feel better. It's a confound with any medicine or treatment but is particularly so in psychiatry which is so often concerned with mood in conditions like depression.

Placebo means 'I shall please' in Latin, and indeed one of several components which constitute the effect may be a desire, conscious or otherwise, for the patient to please the doctor. Other components include spontaneous remission of the symptoms (patients just happen to suddenly get better for reasons unknown), or expectation of good outcome reducing the patient's anxiety.

In very large scale clinical trials of antidepressants the placebo effect was found to be responsible for 75% of the response rate. Remember that that is only for the patients who respond; typically 50% don't respond to the drug at all. So, such drugs are perhaps only helping an eighth of the patients who take them [6].

The placebo effect perhaps explains why there are over 400 types of psychotherapies available today. They can't all be effective in and of themselves.

Disentangling how much the placebo effect comes into play in psychotherapy is very difficult to determine. Double blinded randomised control trials are almost impossible to set up. One example of why this is so can be seen in the case of anti-depressants. Patients with depression are more likely to kill themselves and so, when diagnosed, are quickly put onto an anti-depressant. Putting them onto a sugar pill instead as part of a control study therefore has major moral and ethical implications if some subsequently go on to commit suicide.

The British Royal College of Psychiatrists have said that, based on what few trials there have been, that between a quarter and a third of patients with depression will get better on a placebo (for example a sugar pill) after three months. With real drugs this rises to a half to two thirds. That's not as good as it first sounds because this improvement includes the placebo effect which is bumping up the figures for the 'real' drugs.

Bearing in mind that placebos don't have the side effects of anti-depressants, such as markedly decreased libido in some patients, then there is a lot to say for their use in this condition.

The use of anti-depressants in the young, once widespread, has now reduced following a study which showed that patients were 15 times more likely to actually kill themselves after they had been put on anti-depressant therapy.

As an aside it's worth mentioning that women are much more often prescribed anti-depressants than men. Women are twice as likely to experience depression as men for reasons that are unclear [10]. In the UK, one in nine women seeks help for anxiety and depression at some time. In the US alone around nine million women are considered clinically depressed. Women are nine times more likely to attempt suicide than men. Indeed around 15% of women suffering from severe depression will commit suicide.

This high prevalence explains why there is such a big market for anti-depressants and also why there are approximately thirty different drugs available. These fall roughly into three main types: the older tricyclics (too easy to fatally overdose), monoamine oxidase (rarely used now because of the side effect of high blood pressure) and the currently in vogue selective serotonin re-uptake inhibitors (SSRI) such as Prozac.

The main post war drug to treat anxiety was meprobamate which came to be known as the 'Happy Pill'. Stronger drugs such as valium ('mother's little helper') followed; 2.3 billion pills of this alone were taken a year in the 1960s. Now considered addictive, access to it has become more controlled but during its heyday it stupefied a generation.

Prozac in the 90s became the drug of choice for depression and anxiety, though it was taken by some not to treat depression but to gain happiness. One side effect of the drug was itself depression.

Anti-depressants are known to affect neurotransmitters but there is still no clear evidence to support any of the theories of how they actually work, if they do. As mentioned above, that we do not understand what is going on here is evidenced by the range of conditions, other than depression, for which these drugs are also prescribed. If a single drug can alleviate symptoms in so many conditions then this suggests that the dividing lines between supposedly discrete mental conditions may be false.

It's also worth mentioning that one recent study showed that three thirty-minute walks a week were actually more efficacious in alleviating symptoms of depression than these drugs [13]. Also, just as with the majority of the 'physical' diseases, many of the more common forms of depression will disappear after about eight months without any treatment.

This brings us onto another important point. The World Health Organisation ranks depression as the fourth leading cause of burden amongst all diseases, resulting in an average loss of 12 years of productivity [14]. It also seems to be getting worse and has been predicted to soon become the highest cause of disease burden in the developed world.

But both our Sceptic and Heretic might ask: are all these mood disorders really getting worse because the modern world is somehow making this happen or is there a simpler explanation, namely the changing way that the various editions of DSM have defined them?

Drug medications seem generally to be treating symptoms, but psychoanalysis is trying to cure the condition and often in ways highly tailored to each individual. Determining efficacy of such 'talking treatments' is therefore very difficult because there are no comparable untreated controls.

A more recent 'talking treatment' is Cognitive and Behavioural Therapy (CBT) but it is not regarded as being part of the Freudian psychoanalytic approach because it does not delve deeply into a patient's history.

In CBT attempts are made to modify how patients think and behave and correct what are considered their 'cognitive distortions'. Patients are trained, essentially, to chart a more 'normal' course through life and situations. For example, they may tend to focus too much on negative events and take from this that the world is a terrible place and thus reinforcing what may be a depressive viewpoint. So instead they are trained to take what society generally regards as a more realistic view or are at least given strategies for trying to cope with their perceptions. An illustration may be the use of CBT in patients with stuttering or tinnitus (ringing in the ears). CBT does not reduce the occurrence of either but may reduce the feelings of anxiety or depression the stutterer is prone to.

Despite the difficulties, perhaps CBT is the only 'talking therapy' where real attempts at validation have been attempted, at least to the satisfaction of some private insurance providers if not to the whole psychiatric profession.

In a sense CBT may be comparable to the drug therapies that are used for much mental illness. In essence both are suppressing symptoms but are not attempting to address the underlying disease, if such a thing exists as an actual entity. In contrast, the more psychoanalytical approach assumes that perhaps there is no underlying disease; that the patient's malaise results from the patient's internal world being deranged by external events. By getting the patient to understand these external events they hope to gain a better acceptance and understanding of why they are the way they are, and perhaps even transcend it.

Summary

This chapter has tried to show how deep the problems in psychiatry run. The truth is that we do not know where the mind comes from and have little or no evidence for how it arises from our flesh and blood constituents. Without such basic knowledge it is understandable why psychiatrists have struggled to deal with cases where the equilibrium of the mind appears disturbed. Many of the psychiatric profession have doubts as to the organic existence of many mental diseases, believing instead they are reactions to external events and environment. The contrary view in the profession takes the more usual scientific reductionist and classification approach. When applied to the spectrum of symptoms this has generated more and more possibly fictitious disorders, pathologising many of us overnight.

Numerous drugs and other therapies have been developed to combat mental illness and, whilst some may serve to suppress individual symptoms in some patients, none appear to actually cure.

Both our Believer and Sceptic are ever hopeful that patterns will emerge from the complexity and that we will eventually be able to repair the mind just as we can an automobile.

The Heretic believes that science, its methods and laws and principles, rather than being representative of the bedrock of reality, is actually a product of our limited minds. Attempting to apply mind-generated methods of science to the mind itself, something which transcends these methods, may therefore be an exercise in futility.

In this book we have been travelling from the harder shores of science (maths and physics) to the softer end of the spectrum (biology and medicine). In the next chapter we come to the softest science of all.

12

Social Sciences: Down and Dirty

Sociology begins in the dustbin, and sociologists have always been licensed rag-and-bone men trundling their carts round the backyards of the posher academic establishments.

Alan Bennett, playwright, in 'Writing Home' 1994

He...was a sociologist; he had got into an intellectual muddle early on in life and never managed to get out.

Iris Murdoch, novelist and philosopher 'The Philosopher's Pupil' 1983

The only possible conclusion the social sciences can draw is: some do, some don't.

Ernest Rutherford, physicist.

The social sciences concern themselves with how people relate to each other and the societies they live within. It encompasses sociology, anthropology, education, linguistics, political science and aspects of economics.

Karl Marx is considered as one of the Fathers of Social Science and is certainly the one most people have heard of.

Positive pithy quotes about Social Sciences are hard to come by but negative ones, like the examples above, are much more frequent. The main reason for this is that social sciences are overshadowed by their older, smarter, buffer and bigger brothers in the 'harder' sciences like Physics and Chemistry and Biology.

And for many years I too suffered from the same prejudices of my hard science contemporaries and looked down condescendingly on the 'softer' Social Sciences as not being 'real' science at all. After all, social science experiments were never exactly repeatable and needed to quickly resort to statistics because of the variation in human behaviour that they dealt with. They weren't at all like the experiments in 'hard' science that could be accurately repeated to unearth the bedrock of truth on which a robust theory of the universe could be built.

What it comes down to is that in hard science you can, largely, cut out all sorts of extraneous factors which bedevil every other aspect of real life. If you're studying the relationship between the pressure and the volume of a gas then you can encase the apparatus in a water bath which provides a constant temperature. This is necessary because temperature can affect both the volume and pressure of a gas so you have to remove its effects.

Even in these carefully controlled experiments reality still has the impertinence to intrude. Your constant temperature bath will never be entirely constant and the temperature will never be entirely uniform across the volume. Whatever system you are using to change the volume of the gas (for example a piston) will never be entirely air tight and some gas will leak out and something else will leak in. The sample of gas you use will never be entirely pure and will always be contaminated by other elements.

On top of that, whatever devices you are using to measure pressure and volume and temperature will themselves be subject to myriad sources of error that are a result of their manufacture. Reality always degrades the 'perfect truth' of any experiment.

And finally, you assume all molecules of the gas are the same and behave in exactly the same way. Why this should be the case is not clear; nothing else in the universe seems identical to anything else.

Even so, all the above assumptions usually work as a good approximation.

The trouble with social sciences is that the human beings whose actions and behaviour they study are much bigger than atoms. They are also decidedly different one from the other. Not only that, if you try to 'control' them in order to cut out as many extraneous factors as possible, you immediately start to make them behave in different ways. Shutting someone up in a box to study their behaviour and also to prevent the behaviours and opinions of others affecting that person will only make them behave in much stranger and stranger ways. Not that it's possible to eliminate all the confounding effects that can alter their behaviour anyway.

People in any event are hopelessly non-uniform and messy in how they behave and in what they do and what they say. They're also messy and non-uniform in terms of how truthful they are in telling the experimenter what they've been doing (What? Me drink more than 15 units of alcohol a week? Never!) Even on an individual basis people don't act consistently. One day you might wake up in a bad mood, or feel like wanting to eat a certain food, or watch a certain TV programme or go for a run, but the next day you don't. Physicists aren't troubled by individual atoms doing one thing one day and then, on a whim, doing something different the next. Not that they'd know if that was happening anyway as the damned things are too small.

That social scientists, faced with this maelstrom of uncertainty, nevertheless strive to extract meaning smacks more of heroism than 'intellectual muddle'.

They're certainly more heroic than the 'hard' scientists. Hard scientists are like a politician responsible for a country's defence during war who produces policies (theories) like the Geneva Convention and strict rules of engagement, all based on idealised conceptions of logical and controllable human behaviour. Social scientists are like the soldiers actually fighting hand-to-hand in the trenches.

This book has tried to show that our attempts to understand the universe by hard reductionist scientific methods will always be confounded by its innate complexity. These attempts also pay little heed to the way our brains may be hard wired to think only in certain ways and this ultimately will limit our perception of any real underlying truth. Our lack of success in understanding the mind, for example, may be because we are using our limited minds to study our limited minds.

The limited, subjective nature of our minds and the complexity of systems that can't be broken down into individual bits without totally affecting their behaviour, are both problems that have always been implicit in the work of social scientists. It may be worth briefly considering how they have struggled to deal with these as it may, just may, point a way forward as we get closer to the limits of how far our 'hard' scientific methods can take us.

Social scientists have often been forced by the innate variability and inconsistencies of their subjects to move away from the quantitative (number) based analytical approaches ('positivism') used in the hard sciences towards the qualitative ('naturalist') perspective.

Quantitative studies, such as whether a new drug is effective, can study two large populations: one using the new drug and one using an older drug or just a placebo. The experimenter can use a measure of outcome to compare these populations. If the outcome is to, for example, reduce blood pressure then you have a fairly hard measure.

Another example of this kind of quantitative study would be correlating the amount people smoke and their mortality rates.

This sort of positivist approach actually involves some implicit assumptions which are often taken for granted by hard scientists so are rarely articulated or even recognised. These include:

1) There is only one reality

2) The scientist has to be entirely objective, in other words completely

unbiased, about whatever the outcome is. Inconvenient data won't be summarily dismissed as 'outliers', multiple correlations won't be used to fish out a significant result (essentially, if you compare twenty different things then you will get what is considered a statistically significant result one time in twenty by chance alone)

3) No matter how complex the phenomena, they can be reduced into simpler more readily measurable and analysable units. This reductionist approach means that, for example, the complexity of the anatomy and function of the body can be broken down into countless individual biological processes each of which can be individually analysed. All the analyses can then be added back together to reach a full understanding of the organism. This approach has, as we have seen, singularly failed as far as the workings of the human mind are concerned

4) One thing causes another, the so-called 'deterministic' point of view. Put a flame to some hydrogen gas and it will explode. Cause and effect. [In fact, a contrary view was taken by the 18th century philosopher David Hume with his idea of 'induction' in which the point is made that we cannot be sure of the knowledge that science apparently gives us [1]. His argument was based on the idea that cause and effect may often not be related, it just appears they are. If we repeatedly observe two phenomena in association then we, at some level, will associate cause with effect (a common example quoted is that of white swans. Having seen many white swans one would naturally assume that all swans must be white. It is only when one's experience and knowledge increases, and one sees a black swan for the first time, that that inductive association is broken). Hume applied this reasoning to science and mathematics and suggested that the supposed laws of physics, at least those apparent to us, may themselves vary with time and that what we are seeing may be only a snapshot. However, from this snapshot we construct a comforting model of the universe which could be as erroneous as constructing a whole movie from a single frame of film.

Hume gave a rather unflattering example in which he compared scientists with cattle in a farm. The cattle are fed by the farmer every day, and so the cattle come to associate the farmer with food. They come to believe this association always has been and always will be. They have no concept that they are being fattened for slaughter and that one day this association will catastrophically change]

5) Deductivism. This is the idea that you can test a theory by obtaining and analysing data to determine whether it does or does not support your

hypothesis.

The qualitative or naturalistic perspective is different and uses sometimes diametrically opposite assumptions. It is usually used to look at processes that are not amenable to hard measurements, for example belief systems, behaviours, experiences and attitudes. Hard scientists often recoil from this and indeed this is the source of their disdain for the social sciences. Perhaps you are also tempted to dismiss these nebulous concepts but remember that these are the things that make you what you are. In a sense, can anything be more important?

This qualitative perspective is sometimes called naturalistic because it is often used to interpret and understand human behaviour in a natural setting as opposed to the subjects being isolated and tested in a laboratory somewhere. In such a laboratory setting the subjects are unlikely to act naturally.

The naturalistic approach has certain basic assumptions that reflect the intrusion of messy reality into their research subjects. As you will see they are quite different from the positivist approach [2]:

1) There are as many realities as there are people involved in the research. For example, let's take a hospital. Each person's perspective, their reality, will depend on whether they are a patient, a doctor, a nurse or a relative

2) Reflexive. The researcher is part of the research process and is required to recognise this because they will inevitably affect the experiment. It's interesting that this is a view that most hard scientists would seek to deny unless their field happens to be quantum mechanics. In that case the paradoxical findings of the dual slit experiment have led them, some would say in desperation, to accept that the experimenter is always part of the experiment and not an ideal lofty-minded disinterested, non-interfering observer

3) Ideological. This goes even further than (2) above, in that it recognises that any research is always underpinned by specific beliefs or ideologies. For example, if an ethnologist sets out to study a tribe in Africa he will bring with him many developed world assumptions and beliefs against which he will compare the tribe. For example, how unbiased and objective would you, as a European or North American, be if you were to do a study of female genital mutilation in parts of sub-Saharan and North-East Africa where the practice is both socially acceptable and almost universal. Horrible though the practice may be, could you ever claim to be truly objective about it, because if you had been brought up in those countries you might well have taken a different view?

This is also a problem in hard science though it is not always recognised as such, just as water may not be recognised as an entity by fish as they have spent their whole life immersed in it and can't imagine being immersed in anything else. In hard science this ideology is essentially all the science in the field that has gone before it and which is by default assumed to be correct. If you set out to measure the orbit of an asteroid, you don't start by questioning the 'laws' of gravity (Newtonian or Einsteinian, according to the century you were living in)

4) Holistic, meaning emphasising the importance of the whole and the interdependence of its parts. I must admit that whilst I was in thrall to hard science I used to paraphrase a line in a play by the Nazi poet laureate Hanns Johst: 'When I hear the word holistic I release the safety on my Browning' (the phrase's more commonly recognised form is: 'whenever I hear the word culture I reach for my gun' and has been misattributed to Hermann Goering and also Mark Twain). 'Typical wishy-washy social sciences', I would say dismissively when I heard the word holistic used, mired as I was in my hard science reductionist perspective. But, over the years, I have come to understand that the reductionist approach can only ever take us so far and that, arguably in the biological as well as social sciences, 'holism' may be where the truth lies for everything.

One example of a holistic study might be in studying how a pain killer may alleviate the effects of arthritis by comparing patients given the drug with a control group. Such a more naturalistic study would not simply focus on this but rather explore how the patient was using the painkiller as part of their own overall strategy to manage the pain in their daily lives

5) Inductivism. This is the opposite of deductivism, where one starts with a theory then tests to see whether it is true or not. Instead one develops the theory from observing the subjects. An example of this is the work of Elisabeth Kubler-Ross [3] who studied the experiences of people who had been told they were dying from terminal illnesses. These observations led her to identify five stages through which the patient went, namely denial ('it must be some mistake'), anger ('why me?'), bargaining ('if you let me live God, I'll never bet on the horses again'), depression and finally acceptance. Ross didn't start out with a theory she then tested. Even though this is obviously a broad-brush stroke finding with considerable variation, few would doubt an element of truth in what she found.

Indeed, our Heretic, convinced there are no underlying principles or laws to the universe, would say this inductivist approach is the only way to study the universe; to look for examples of where there is presently some form of repetition or regularity that may be useful.

Quantitative researchers design studies; qualitative researchers talk about research approaches.

Quantitative research designs can look at changes in a measurable parameter for a population at a given time point (cross-sectional or 'snapshot' studies), or how it changes with time. Randomised control trials can be performed. Data can be acquired prospectively or data already acquired can be used retrospectively.

Social scientists and other 'softer' researchers often have to plot a different course. They might like to take the quantitative approach but subjectivity is more manifestly present right from the start than it is in quantitative studies. Often when studying feelings and behaviour, not in the lab but out in the real world where one often has to rely on reported behaviour by the subject, then questionnaires have often to be employed. The decisions as to what questions will be asked and how the answers are to be interpreted allow subjectivity by the experimenter to creep in.

Qualitative researchers often have no choice but to use descriptive or exploratory approaches. They may attempt to understand what is happening in a given experimental situation from the perspective of the person involved, for example that of a patient being treated for an illness. Data comes from interviews or from small 'focus group's (despite myself, I still can't help wanting to release the safety on my Browning when I hear that phrase). These are where small groups of subjects may serve to tease out common feelings and the way they understand their experiences [4].

From these initial explorations, themes can be generated. Let's take for example an attempt to understand different cultures. This doesn't necessarily mean putting on our boots and stumping through a rain forest to study a tribe deep in the Amazon; perhaps instead the researcher wants to study a group of elderly people in a care home. Like a tribe in a rain forest, or stock brokers in the City of London, they will have developed their own local culture. Every human group eventually evolves a culture with its own world view and also its own 'tacit knowledge', that's knowledge so embedded in the culture that it is not talked about by the group and indeed they may not even be consciously aware of it at all. It is said that the reason reality TV shows such as Big Brother can attract large audiences is that watching peoples' behaviour so closely can reveal deeper insights into the tacit assumptions the viewers themselves make in their own lives.

Whether the researcher is studying the experiences, beliefs or behaviours of the head hunters of Borneo, or of student nurses in their first year of studies, they have to decide whether to take the insider's (emic) perspective and essentially 'go native' by actually joining the group, or to hang onto the shreds of objectivity and take the outsider's (etic) perspective.

Many social scientists also take the approach of trying to produce grounded theories. These are studies where data are collected, often in the form of interviews or questionnaires, and only after that is a theory constructed grounded on this data. Even reviewing the previous literature on the subject should be left until after the data is acquired so as not to bias the observer. They do this because the researcher is trying to avoid any of their own or others' preconceptions about the area. This approach is contrary to the hypothesis driven experiments of hard science.

Once a theory has emerged by this process then further targeted subject groups can be investigated to develop the emerging theory. Data collection and analysis occur together, and not one after the other as in quantitative research.

Hard scientists rebel at what they would consider such nebulous and often implicitly subjective approaches to research. But the fact is that these softer scientists live in the real world where reality is messy and essentially uncontrollable and where the impossibility of being objective is acknowledged.

In fact, it is the hard sciences that are the unnatural ones, where experimental conditions are constrained to within an inch of their lives to allow the God-like objective scientists to study one specific effect in isolation; a situation which never pertains in the living, working universe.

Perhaps one day we may have to give up the search for 'objective' truth, something that may itself be an illusion of our limited minds, and instead develop new ways of thinking about the universe. Perhaps when that day comes the hard sciences will look to the social sciences for guidance.

13

Long Timescale Controversies: Evolution and Climate Change

It is absolutely safe to say that if you meet somebody who claims not to believe in evolution, that person is ignorant, stupid or insane (or wicked, but I'd rather not consider that).

Richard Dawkins, Evolutionary Biologist

Why, these men would destroy the Bible on evidence that would not convict a habitual criminal of a misdemeanor. They found a tooth in a sand pit in Nebraska with no other bones about it, and from that one tooth decided that it was the remains of the missing link. They have queer ideas about age too. They find a fossil and when they are asked how old it is they say they can't tell without knowing what rock it was in, and when they are asked how old the rock is they say they can't tell unless they know how old the fossil is.

William Jennings Bryan, Democratic politician and Christian

We have reason to believe that man first walked upright to free his hands for masturbation.

Lily Tomlin, comedienne

All across the world, in every kind of environment and region known to man, increasingly dangerous weather patterns and devastating storms are abruptly putting an end to the long-running debate over whether or not climate change is real. Not only is it real, it's here, and its effects are giving rise to a frighteningly new global phenomenon: the man-made natural disaster.

Barack Obama, speech, Apr. 3, 2006

With all of the hysteria, all of the fear, all of the phony science, could it be that man-made global warming is the greatest hoax ever perpetrated on the American people? It sure sounds like it.

James M. Inhofe, Republican politician

President Bush said global warming is happening much quicker than he thought, and then his staff pulled him aside and said 'It's just springtime'.

Jay Leno, talk show host

Both these subjects have provoked bitter polarisation to the point where each side claims conspiracy and fabrication on the part of the other. If someone shows even the slightest scepticism for either position, then they are immediately viewed as a proselytiser for the opposing camp. Zealotry leaves little room for shades of grey.

In this book we've been taking a hard look at science and have made much of its limitations and assumptions. Here we're going to take at least a sceptical view of the evidence for these two current subjects of controversy. Both are linked by our limited capacity to accurately determine what happened in the distant past.

Bearing in mind the polarisation that these topics provoke, let me state my own circumstances so we're all clear:

I have never received funding from the energy industry nor do I know anyone employed by it.

I'm not a Christian. At best I'm an agnostic about all religions, though on a bad day I may even stray into outright atheism.

I am at least sceptical or even heretical about science and its methods.

In short, I am not a fan of any of the sides in the debates about evolution and climate change.

I therefore maintain that I have no axe to grind on either issue.

For most of the science discussed so far, one can make measurements to establish the value of something; for example, the force of gravity or the rate of a chemical reaction. One measurement is made at one particular time point. In some cases, these measurements have been repeated over several hundred years. The estimation of the force of gravity is one such example.

It's true that the further back you go in time the less accurate these measures are likely to have been due to technological limitations in the measuring equipment of the day. Nonetheless, experimental evidence indicates that gravity has been more or less constant over that time period. This leads to the assumption that gravitational force is a constant that does not vary over time. Without accurate measurements over much longer timescales, we can't be sure the assumption is correct but until contrary evidence arises it is a reasonable one to make.

But what about the situation where what we are attempting to measure is known to have changed markedly over time? If mankind has only been able to make accurate physical measurements over a few hundred years, then how can we determine changes that have taken place over thousands of years or even hundreds of millions?

This uncertainty is why the issues of climate change and evolution are fertile grounds for vitriolic controversy. If these long-term changes were limited to some specialised esoteric field, like the mating habits of the dormouse, the controversy would be limited to a few academics. However, when it comes to climate change and evolution, hulking commercial and religious agendas serve to drag these issues before a much wider audience.

Modern civilisation has developed because it has been able to harness the power from fossil fuels: the oil and gas and coal that lie under our feet. The majority of climate scientists presently believe that mankind's burning of these fuels has caused an increase in carbon dioxide levels, resulting in a 'greenhouse effect' and increasing global temperatures.

If this temperature rise becomes significant, and prevailing wisdom is that it will do, then ice caps will melt and sea levels will rise. Populous coastal cities will be inundated. There will be severe droughts and famines; hundreds of millions of people will die or be displaced.

On the other hand, reducing fossil fuel usage would compromise economic growth. Indeed, how could our modern world function if we couldn't even heat our homes? How would we feed our large populations when we couldn't use road or sea or air transport to get the food to our tables?

Alternative means of power generation, such as wind and tide and solar, are desperately being improved but still produce electricity that is generally expensive or too inconsistent compared to that produced by burning fossil fuels.

It would perhaps ultimately come down to a choice between a reduced standard of living for the developed and developing worlds or damnation for the global poor.

Everyone, it would appear, has a stake in knowing if the science of man-induced climate change is correct.

The concept of biological evolution, as first described by Charles Darwin, is one that provides a theory that tries to explain the biological diversity of life on the planet. It does so by requiring small incremental change over millions of years. This conflicts immediately with the views of some fundamentalist Christians that the Earth was only created within the last 6000 years. They believe that fossils, and other evidence from before that time, are the results of a cataclysmic flood that wiped out all life-forms.

There is another, though less fundamental, Christian view based on creation science that has it that God made all the different species. That they therefore did not all evolve from one organism, as Darwinian theory would have us believe.

Fundamentalists, and indeed many other Christians, feel unease at how Darwinian theory essentially reduces the role of God in the biological development of man. It suggests that the most that God did was light the blue touch paper and then retire.

Both evolution and climate change are events that occur over long time scales of thousands or millions of years. Both therefore immediately butt up against this measurement problem. Without hard direct evidence, such as measuring the daily temperature for a few million years, or studying a particular species over even longer timescales, then there will always be fertile room for doubt about what has actually happened.

Whilst scientists cannot provide direct measurements over these timescales, they can take surrogate (indirect) measurements such as examinations of Antarctic ice cores for evidence of climate change, or of the fossil record when it comes to studying evolution.

However, as soon as we move away from direct measurement then theory immediately becomes involved. If history teaches us anything it is that theories never last. They too change with time.

In this chapter we will look, with a sceptical and sometimes heretical eye, at the theory behind the surrogate measures we use to assess climate change and explain evolution.

The first question is: how do we date these things?

Dating Techniques

[Spoiler alert: this is one more case of how, when looking closely at any technique or simple theory, it soon starts to break down]

The most well-known of the methods for ageing samples is carbon dating. It can only be used on carbon containing compounds such as wood and clothing and is based on the radioactive decay of the unstable isotope carbon-14, with a half-life of 5730 years, to the stable carbon-12. In other words, over nearly six thousand years, half of all carbon-14 atoms will change to carbon-12.

Supposedly accurate up to about 60,000 years, there are known to be a number of potential confounds that serve to reduce its accuracy. The following describe only those considered most significant:

1) The assumption is made that the atmosphere, from which the carbon for organic molecules comes, has maintained a constant proportion of carbon-14 to carbon-12. There is some evidence from samples in tree rings that this ratio has not always been constant. The reasons for this could be due to older, stable carbon being released by melting glaciers and, more latterly, by man and his burning of fossil fuels. Changes from tree ring to tree ring that are the equivalent of year to year changes, are used to try to calibrate out this effect.

2) That the rate of radioactive decay has remained constant over time. If it didn't that would put a major spanner in the works as far as physics is concerned. Measurements of radioactive decay have only really been performed for little more than a hundred years. They've been constant within experimental measurement over that time. Until we find experimental evidence to the contrary it seems reasonable to assume this is correct. It is worth noting however that in the time of the inventor of the carbon dating technique (Willard Libby in the 1940s), the half-life was measured at 5568 years. Later the measurements were 'improved' to give a figure of 5730. The likeliest explanation for the discrepancy, and the apparent change in half-life values with time, is simply measurement error. Objects containing organic molecules such as... well...you yourself, are hardly buzzing with radioactivity and it is difficult to measure these small decay events against the background of radiation from the stars (cosmic rays) and from radioactive elements in the earth itself such as uranium and radon. Nevertheless, the idea that the half-life does not change over time is definitely an assumption.

3) Carbon-14 is produced by cosmic rays interacting with nitrogen atoms. Two more assumptions are therefore made: that both cosmic ray activity and the amount of nitrogen in the atmosphere have remained constant over the last 60,000 years or so. There are certainly factors that can affect the former. Oscillation in the Earth's magnetic field can alter the number of cosmic rays striking the atmosphere; the higher the field the more rays are deflected which in turn can alter the amounts of C-14 created. These oscillations happen over intervals of around 10,000 years. C-14 released from extensive atmospheric testing of nuclear weapons in the 1960s is another confound. Additionally, sunspot activity is variable and so produces varying levels of the cosmic rays. These again, hopefully, are calibrated out using samples from

consecutive tree rings.

4) Carbon from plants is the source of most of the dating. Such plants may be used to make clothing or they may be ingested by animals and be incorporated into their bodies. The trouble is that plants do not take up C-12 and C-14 in equal quantities (this is called isotopic fractionation). Though all plants favour the lighter isotope, they do not do so at the same rate. There is even wider variation between land and water based photosynthetic organisms such as algae.

5) Volcanoes. These spew out carbon so old it has hardly any C-14 atoms left. Carbon dating of local plant life whose life-spans are actually only a few years can give erroneous carbon based ageing figures of many hundreds of years.

6) Dating samples like bone is prone to many confounds including contamination during its collection but also by the loss of carbon carried away by water in the ground where it's buried (the carbon is not found in the bone itself but in the marrow spaces within the bone through which the water may percolate). This carbon loss is also a temperature dependent process so environmental conditions can affect the accuracy of the measurements.

Care must also be taken in terms of where the organic material originally got its carbon from. For example, if the carbon came from water from underground streams this may have percolated for hundreds or thousands of years through rock before being ingested. On its way it may also have picked up carbon from limestone (calcium carbonate) that is many millions of years old. This is called the 'hard water' effect.

Carbon in sea water can also be a very big problem. For example, if you were to carbon date the fish you had for tea last night you would find it would probably appear to be around 400 years old.

Perhaps you should change your fishmonger.

In fact, this is because sea water, and the carbon in it that came originally from the atmosphere, can circulate for thousands of years in the deep ocean before coming back to the surface and being 'breathed' by the gills of our unfortunate fish [1].

So, carbon dating is subject to a number of often quite significant errors.

But aren't there objects of known age we could compare carbon dating results with to see if the technique is accurate? Indeed, the carbon dating technique was originally validated by analysing wood from Egyptian tombs whose age had been dated in the historical record to around 2625 BC, give or take about 75 years. Carbon dating gave a figure of 2800 BC, give or take 250 years.

Not bad, but can we prove it is accurate further back than that? Due to the paucity of the historical records beyond this point, we can't.

In any case, for evolutionary studies that involve processes lasting millions of years, measurements of which are usually based on fossils which no longer contain any carbon, then we need some other technique. For this, Radiometric Dating is our man.

This dates rock but does so again by using the ratios of radioisotopes that occur in the rock and their decay products. Half-lives of the isotopes used can be as long as 100 billion years as the decay rate is so small. So small in fact that it is difficult to measure accurately.

On the other hand, these rocks will generally have been buried for much of their time and so measurements of their radioactivity will be much less affected by any variations in cosmic rays, or by the radioactive fallout from atmospheric testing of nuclear weapons, than the organic samples used for carbon dating.

The technique is however still subject to the assumption that the half-lives of these isotopes have remained constant over the millions and billions of years since the rocks were formed.

It is also possible that water permeating through the rock will have changed its constituents, preferentially carrying away some of the radioactive decay products. Similarly, heat can cause some of the isotopes to differentially diffuse out.

Despite this, margins of error on uranium-to-lead based dating are said to be less than one percent for rocks that are over a billion years old.

Even our Heretic would be hard pressed to reject the general truth of radiometric dating. It is after all simply exploiting regularities in the behaviour of the universe. Half-lives and decay rates may be mathematical forms we have applied to the process, and the limits of our theory are laid bare by the fact we can never predict when a given atom will decay, but nonetheless we can with reasonable accuracy predict the decay of large numbers of atoms.

The Heretic may grumble that we don't know for sure that half-lives have remained constant over time but, even if there were such a variation, would rocks we measure to be 4 billion years old really only be 6000 years old as some fundamentalist Christians might claim?

Incidentally, some fundamentalists get around this problem by saying that God used existing materials that could have been billions of years old to create the Earth, that the Earth was wiped clean 6000 years ago and life started again. In other words, there was an apocalyptic reboot to give us version 2.0. That would also explain how we get such old fossils as they actually came from version 1.0.

But, like the dating techniques, there are a lot of assumptions there. Not only that, but the only evidence for this creationist view is provided by the Old Testament. If we're being hard edged enough to be heretical about science and questioning even theory based on at least some repeatable experimental evidence, then we can't accept the Old Testament as fact. That, in a sense, is theory based on no repeatable evidence at all.

As for climate change data, ice cores from the Antarctic are the main means of establishing atmospheric carbon dioxide levels back to about 800,000 years ago. Each new snowfall traps a layer of air. The further in the past this happened, the deeper this lies in any ice core that is drilled out of the ice. The depth in the core can be matched with age by isotope dating and so is subject to some of the errors discussed earlier. Over and above those errors, the amount of trapped air can vary depending on how quickly a consolidating layer can be formed over it. Otherwise during consolidation older, lower lying gas can diffuse up into younger overlying years, thus causing erroneous estimates of age. This source of error alone can cause mis-dating by up to 6000 years.

It's not impossible that all our dating techniques for material over 10,000 years old may be wildly inaccurate but there is at the moment nothing to suggest they are. In the following discussion of the two big controversies we will assume that despite the errors techniques such as dating of fossils for evolutionary purposes, and ice cores in Antarctica for estimating climate change measures, can be used as at least rough indicators of age.

Evolution

So radiometric dating tells us the Earth is really old but can it be possible all life arose with each species fully formed only 6000 years ago as some Christian groups would interpret the Bible as indicating? As we have seen there are many assumptions in carbon dating and many sources of error but, even so, the technique does seem able to date human and animal samples and human artefacts back to about 60,000 years, the limit of the technique. Sure, there are innate errors but enough to reduce 60,000 to 6,000? On balance, taking the Sceptical or even the Heretical points of view, one would be hard pressed to think that likely.

Charles Darwin in his book On the Origin of the Species by Means of Natural Selection, or the Preservation of the Favoured Races in the Struggle for Life published in 1859 tried to explain the diversity of life on Earth without resorting to the idea of God designing every species separately. Darwin believed that one species could change into another over time. He proposed that random mutations produced small changes in an organism. If the environment suddenly changed then that new mutated creature might have an advantage over other members of its species in terms of survival. For example, a mutation that allowed a rodent to see slightly better at night would be good news if the population of night hunting predators suddenly increased. This mutation could then be passed onto the mutant's offspring while all the 'normal' rodents were being eaten. Over time only the mutated rodents would be left.

Mutation was seen as a random event and the vast majority of mutations would be fatal or would make the organism less likely to survive. Only the lucky mutations that better suited some environmental change would be transmitted from generation to generation. Many mutations would, however, be required to change one species to another.

Darwin's theory is one explanation for the diversity of life on Earth but it isn't the only one. Indeed, right from the start there was a problem with some of Darwin's ideas, and a swift re-writing was necessary to produce the current orthodoxy of Neo-Darwinism. As well as his survival of the fittest route to evolution that is still a tenet of Neo-Darwinism, Darwin believed in the inheritance of adaptations made to the bodies of the parents in their own lifetimes. In other words, if you learnt to juggle before you had kids, you might then pass that skill on to the kids genetically. Neo-Darwinism, the current prevailing orthodoxy, rejects this idea of inheritance by experience. In other words, they reject the idea that genes are altered in an organism's lifetime and can then be passed on in a similarly altered form to the offspring.

Even with this change to the old theory, Neo-Darwinism still has major problems of its own, as we shall soon see.

Perhaps even the Sceptic must wonder whether Darwinian survival of the fittest is really the only, or indeed even the most important, driver of biological change for life on Earth [see [2] and [3] for more in-depth discussion of this]. Each biological system is just so overwhelmingly complex, as was illustrated in the chapter on genetic transmission of information, that one wonders how a simple survival of the fittest mechanism could provide it. How many environmental changes would it take to produce DNA strands containing over 200 million molecular units, or the million different proteins or the more than a billion different species that have at some time existed on the Earth?

At this point we'd better define exactly what a species is. Bankers and Estate Agents and Loan Sharks are not, as some members of society might hold, separate species to the rest of mankind. That is because the rest of us could breed with them if it was absolutely necessary. Lions and Great Danes, on the other hand, cannot breed with each other no matter how hard they try because they are separate species. Wolves, however, could mate with Great Danes because dogs of whatever flavour are sub-species of wolves, though it would be advisable not to be in the vicinity when this happened.

It's worth pointing out that categorisation into species is, as with any scientific categorisation, never as clear cut or absolute as it looks. To quote the American biologist E.O. Wilson on the definition of species being a population that can breed freely and naturally, such a definition 'is filled with exceptions and difficulties reflective of the complexity in evolutionary biology'.

Nevertheless, separation into different species clearly marks a big change because it means the new version can no longer breed with the previous one.

Neo-Darwinian theory essentially says that the recipe for life is based on genes. Genes cover everything from eye colour to susceptibility to disease. These genes can spontaneously change, in other words mutate, because of chemical changes in the DNA molecule.

These changes can initially cause 'vertical evolution'. For example, the members of a species of butterfly may start to differ from each other by the colour of their wings though that will not stop them interbreeding.

Factored into this is environmental change suddenly giving an advantage to one of the new mutations. The commonly used example is that of the giraffe. These may originally have had short necks, though a mutation gave some of them longer ones. Whilst relatively short trees were available this didn't matter, but if the short trees were killed off by disease then only the longer necked giraffes would survive.

That's apparently a good and clear example but it may also hint at some limits of the Neo-Darwinist view. Indeed, some would say it is actually an example for the counter argument. Such a change in neck length by itself would not be enough and indeed would be fatal to any giraffe. For example, if a human was born with such a long neck and tried to breathe through this eight-foot-long tube, they would perish. Not because of a lack of oxygen but because they would be constantly breathing out then breathing back in the carbon dioxide that was being produced by their respiration.

To avoid the same thing happening to the new long necked giraffe, profound changes were also required to its cardiovascular system and also to allow blood to be pumped against gravity up to the brain through the long neck. This produced such a high blood pressure that it would drive blood out of the capillaries in the lower legs of the animal. Additional changes to the skin had to happen at the same time to prevent the high-pressure fluid in the spaces between the cells from leaking out of the skin.

That all these quite profound changes happened fortuitously and simultaneously seems highly unlikely. Even after them, though, these suddenly re-engineered giraffes would still be part of the same species.

But even such profound changes are therefore nothing compared to those required when one species turns into another entirely. Countless mutational changes would be necessary then.

One concern with this process called speciation is that it has never been observed. This is despite many experiments over a century on countless millions of organisms with rapid generational turnover such as fruit flies. These can produce a new generation in less than two weeks. Though new mutations have been observed to become prevalent amongst these populations, thus providing some support to Darwin's idea of vertical evolution, mutations have never accumulated in sufficient numbers to actually create a new species.

And that's a problem because of the sheer number of species there are. It has been estimated that there are presently 1.4 million known species of animals, plants and micro-organisms on Earth and we're discovering new ones all the time. It has also been estimated that those we know of represent perhaps only about a tenth of the species presently living on the planet. So, there may be something like 14 million different species on the planet today.

That's amazing, but it gets even more stunning when you include the estimate that this only amounts to 1% of the species that have ever existed. This means that perhaps 1.4 billion different species have trodden, flown, put roots down in or squirmed over the Earth at some point. And that's the lower estimate. Some put the figure as high as 16 billion [5].

Can such mind-boggling diversity be explained by the simple Darwinian model?

Not only is there such a plethora or species, but one's mind also reels at the complexity within each cell of each organism. On the surface of each cell are millions of openings configured in many different ways to allow only certain molecules in. From this surface, myriad conduits branch away to take these molecules to the nucleus or other organelles in the cell where they are then processed to make other molecules. Inside the nucleus are the 220 million-unit strands of DNA, with molecules shuttling to and from this to other parts of the cell. Many of these molecules are proteins, of which there are tens of thousands of different types, each with perhaps 3000 atoms all folded in highly complex and unique 3D shapes.

Could this mind-blowing complexity really have resulted from a series of environmental changes? How many changes could there be on just one planet, four billion years old though it seems to be?

Even Darwin said: ' I am convinced that natural selection has been the main but not the exclusive means of modification'. Many respectable scientists go even further nowadays and believe selection of the fittest isn't even the main mechanism of evolution [2].

There are many camps of informed doubters. Some, the 'patterned' or 'transformed cladists', whilst accepting there has been evolutionary change, and agreeing that a species' bodily structure and functions can undoubtedly change (vertical evolution), are particularly excited by there being no hard evidence of even one species ever changing into another. They are convinced this shows there must be other mechanisms at work that govern the behaviour of complex systems and cause species transformations.

Others take the older, and long since dismissed, Lamarkian view that agents in the environment change an organism. Additionally, and more importantly, somehow these changes are passed to the organism's offspring. In other words, 'nurture' rather than the straightforward Darwinian 'nature' is another source of genetic material.

Some put forward other mechanisms for evolution [4]. They point out that DNA contains information that codes for many different proteins and their subsequent structure and function, but very often whole sections of DNA just aren't activated. Often these sections are dismissed as 'junk' DNA though it may simply be that we do not yet know what they are actually doing. One possible function of this 'junk' DNA is enabling the organism to show genetic flexibility. Environmental change may somehow activate more appropriate sections of the DNA.

Examples such as the peppered moth are often cited to show natural selection in action, though it is actually an example that shows quite the contrary and may support this alternative idea. Pre-Industrial Revolution peppered moths were white and were therefore camouflaged against the white lichen that covered trees. However, industrial pollution saw off much of this lichen and made the white moths more conspicuous and so better targets for predators. However, the population of moths was not, and had not, always been white; occasional black ones were also always produced. These black moths were the ones to survive by natural selection when the lichen disappeared, but the important point to bear in mind is that a new gene was not created by the process of natural selection; it already existed. Rather than a genetic advancement it was, in fact, a sideways step. The point is that species already have some flexibility built into them.

As well as the overwhelming variety of species that exist and have existed on the planet, causing many to wonder whether environmental pressures could really cause all this genetic change, we also face difficulties posed by complex structures such as the eye or the wing. Did a whole wing just suddenly evolve with sufficient strength to allow flight? That would hardly be possible under existing classical Darwinian theory where there would have to have been many intermediate steps and mutations. Intermediate stubby wings not sufficient to allow flight are more of an encumbrance than a help and would make the organism even more likely to be eaten by predators.

In fact, there is an actual example of a bird whose wings weren't capable of flight: the Dodo. How did that work out?

To fly, you need more than just wings. Feathers help, as do much lighter bones. A faster metabolism and keener sight are also necessary. Whilst evolving though all the classically necessary intermediate stages, proto-birds would have been stumbling around more slowly than others in the population, encumbered with as yet useless stubby wing appendages, with half grown feathers and bones still too heavy to allow flight. Why weren't they all eaten before they managed to take to the air to escape the ground-based predators that were presumably one of the 'evolutionary pressures' that drove the development of wings in the first place?

There are also examples of where evolution seems to have happened without environmental pressure [5]. For example, squids developed water jets before there was anything around to use them to dart at or away from. Environments such as the Great Lakes of the Rift Valley appear to have been very stable over time but have hundreds of different kinds of fish in them. If there was no environmental change to drive evolution then why should there be such diversity?

Another alternative idea is of some mechanism in the DNA which seeks to experiment; that it is not always random chemical or radiation effects that cause mutation. Such a mechanism might work with or without environmental pressures. What these pressures would provide in fact would be the testing stages for new designs. If there was something implicit in the genome that made it try out all sorts of modifications, then that might explain the burgeoning species we find even in stable environments like the Rift Valley lakes and would give flexibility to the process of evolution.

A process which automatically throws out variations might also better describe how bird wings might develop in the sense that change would be much quicker. There would be no need to wait for a series of infrequent environmental changes to favour certain mutations.

It would also help explain why the similarly complex structure of the eye has developed independently in different lines of species at least six times (Even Darwin said: 'When I think of the eye I shudder' [5]). Even happening once by sheer Darwinian chance seems to be pushing the bounds of likelihood.

Such a driver of evolutionary change could reside in the genome itself. Not only do some genes code for specific proteins, but others seem involved in the process of activation. In other words, they govern when certain genes are turned on and off. Perhaps some external event, possibly linked to environmental change, may trigger the DNA to activate new proteins which alter the structure and function of the organism so it can better cope with the change precipitating this. Indeed, this idea may explain why after a hundred million years or so of mammals basically going nowhere in an evolutionary sense, then suddenly and across at least a dozen species, they began to show the same changes. For example, the way mole-like creatures appeared in America and Asia at the same time, or anteaters in four places too far spaced to have been colonised so quickly. These simultaneous transformations happening by the chance mutation is not likely to have happened across all of them at the same time.

This possible mechanism may also explain why so few intermediate states can be found in the fossil record as one species supposedly transitions to a new one. Instead, new sections of DNA may be being suddenly activated, new proteins being produced which engender immediate changes in structure.

Such a theory does not explain where this catalogue of unactivated genes came from so let's not get too excited about it. I'm mentioning it just to show that alternative explanations for evolution are around.

This alternative theory also throws up a disconcerting oddity which makes one raise an eyebrow: the more DNA an organism has, then presumably the more potential it has to evolve. That means the organism most likely to win the evolution states, the one with the flexibility to evolve the most is none other than (drum roll)...the lily.

Even Darwin would have agreed that something else is going on here. In Chapter 5 of On the Origin of the Species Darwin says he is inclined '...to lay less weight on the direct action of surrounding conditions, than on a tendency to vary, due to causes of which we are quite ignorant'. However, Neo-Darwinists have moved away from this and become dogmatic about survival of the fittest being essentially the only mechanism.

Changing from one species to another is difficult enough to explain but there are even greater and yet very sudden changes that have happened. For example, the changes from fish to air breathing creatures with four legs or, going even further back, from brainless, boneless jellyfish to fish with spines and brains.

And there are many other examples of sudden massive changes. Whales and dolphins appear in the fossil record fully formed and diversified. Eggs are similarly found fully developed as indeed are feathers. Rodents appeared suddenly equipped with specialised gnawing teeth. Fish seem to have come out of nowhere as no direct ancestors can be found. Mammals seemed to have appeared about 200 million years ago and then didn't seem to change much until 75 million years ago. At that point and over a period of only 12 million years, they diversified into numerous quite different forms such as marsupials, rodents, camels, carnivores, marsupials, whales, dolphins, rabbits, bats and many others. What happened to animals is only one example of what palaeontologists call 'explosive radiation' with fishes and reptiles showing other examples.

Evolutionists call these sudden discontinuous changes 'saltations' which means jumps. Darwin himself was puzzled by these macro-evolutions and asked: 'Where are the infinitely numerous transitional links?' that the sort of micro-evolution mechanism he was proposing would require [5].

One explanation might be changes in the rate of mutations. This should be constant over time according to Neo-Darwinism but perhaps it isn't constant at all. Perhaps at the time of these discontinuities it sped up beyond recognition for reasons unknown. Neo-Darwinism certainly does not explain these sudden fits and starts.

There are other aspects that do not fit with Neo-Darwinist orthodoxy. If it was true then one would assume that faster evolving types of animals or plants would survive longer as they could more quickly change to suit changing environmental circumstances. This doesn't seem to be the case with oysters and mussels that have changed little over 400 million years and yet are still with us. As indeed are 'living fossils' like coelocanths, opposoms, and king-crabs.

Neo-Darwinism believes that rapid evolution is brought about by three things: a large population, a rapid mutation rate and a short generational turnover. All these factors would increase the probability of throwing up better adapted mutations to any change in environment.

There is however one species that has changed very rapidly but has none of these features. That creature is us. Indeed, the population of man worldwide in Stone Age times was perhaps only in the hundreds of thousands. How come so few of us, with our long generational turnovers due to the many years required for us to come to sexual maturity, managed to change so much over so short a time?

It's also interesting that microbes, to which all three of the above conditions apply, and of which there are countless millions in just about every single animal, have evolved so little over billions of years.

None of this is exactly evidence of some other unidentified but strong evolutionary mechanism at work but it is suggestive of it.

Neo-Darwinism is the modern take on Darwinism and is definitely the accepted wisdom of the age. Its adherents are very sensitive to criticisms of Darwinism as nowadays they always suspect it may be an attempt at forming a beachhead by the promoters of Intelligent Design. This is the belief that

God basically designed all life and is responsible for any changes, though of course there is no proof for this.

Unfortunately, this polarisation prevents the Neo-Darwinists from really considering that natural selection may only be a small part of the evolutionary process. In doing so they have, in effect, painted themselves into a corner.

It is worth pointing out again that no scientific dogma ever lasts. Practitioners in the medical sciences believed in the 'humours' for nearly two thousand years. Until Einstein came along, physicists almost universally believed that Newton had gravity more or less taped. Until about 1990 cosmologists believed that the expanding universe would slow down and agonised only over whether there was enough mass to make it all collapse into a Big Crunch. Now experiments indicate that the expansion of the universe is actually accelerating.

The message from this to budding heretics out there is not to be intimidated when the vast majority of scientists believe a particular thing at a given time. If history has taught us anything, it is that they are always wrong.

It is equally worth emphasising that just because Neo-Darwinists may be wrong, this does not mean that Intelligent Design is right. There is absolutely no hard evidence to support that.

The conclusion, that at least our Heretic would come to, is that scientists are using a single simple 'Law' to explain a complex system. Rather than explaining things, this actually throws up more questions than answers. Physics has had to invent a plethora of laws to explain the burgeoning complexity of the workings of the universe, and even then, not very successfully. Some say Darwinian evolution is the one great law of biology that is required even though it produces a complexity of form and structure that makes the rest of the universe look positively uniform and simplistic by comparison.

The Sceptic might well accept that Darwinism isn't the whole story and might not be surprised that other driving forces for evolutionary change will be discovered in time. That may take a while as, driven into an extreme position to try to counterbalance the forces of religious fundamentalism, evolutionists are not exactly in a 'good place' to even acknowledge the limitations in what their present theory can explain. Without that acknowledgement by the scientific establishment, few will have the courage to seek alternative explanations.

The Heretic might still doubt that the 'discovery' of other laws of evolution will help in the long run as he doesn't believe there are any such laws in the first place. That said, he's very much on the back foot to explain why a universe that 'just is' would somehow produce such rich diversity of organisms as well as the overwhelmingly complexity in the cells that constitute them. He's even forced to grudgingly consider that maybe there is at least one law under-lying the whole thing; a law that forces everything in the universe to greater and greater complexity and diversity.

Climate Change

Carbon dioxide and water vapour act like a blanket over the Earth, ab-sorbing and re-emitting radiation. Together it has been estimated they make a 21o C difference in the temperature of the Earth. With this blanket effect the average surface temperature is about 15o C. Without it, temperature would be -6o C, and that would make it very difficult to sustain human life as we know it.

Water vapour is more important as a greenhouse gas than carbon dioxide but, strangely, climatologists don't worry about that as it is not believed to be changing due to human activities.

I personally began to harbour doubts about the received wisdom on climate change a few years back when there was a little mentioned shifting of the goalposts. Even back in the 1990s a large consensus was saying that mankind was inducing global warning because of the huge amounts of fossil fuels we were burning, thus increasing the amount of carbon dioxide in the atmosphere and increasing the 'greenhouse effect' brought about by the car-bon dioxide blanket. In the 1990s such climate predictions promised warmer, wetter winters because of this effect.

This tune changed however when the UK and Europe were surprised by a couple of very severe winters. In order to draw this errant observation into the fold of theory, climatologists began to say that global warming was mak-ing the ice caps melt faster and this extra cold water was compromising the Gulf Stream, nowadays more commonly referred to as the 'Atlantic conveyor'. This surface ocean current brings heat from the tropics to the western shores of colder northern latitudes.

Suddenly colder winters were being touted as evidence of global warming. That kind of volte face troubled me as this suggested a theory that was neither comprehensive nor firmly grounded.

Not that I worried either way. I've lived for long enough in the UK to know just how variable the weather can be. Not being in the middle of a con-tinental landmass, the UK doesn't have hot and humid summers or very cold and prolonged winters. Instead it is at the mercy of what weather the Atlantic brings. Making inferences from weather over a few tens of years is meaning-less for the UK at least.

And, really, that's the problem. Climate, as we will see, varies all by itself and far more catastrophically than even in those few bad winters we saw.

The whole theory of global climate seems deceptively simple at first. It's a balance, surely, of what energy comes to the Earth and what energy it gives out? However, as with anything, if you scratch the surface of the theory you find vast interlocking complexities even before you put the fossil fuel burning of man into the pot.

And complexity is, as we've seen, not something our linear brains or the linear processes of our computers can readily handle.

Let's consider some of the confounding factors that serve to undermine our simple climate predictions.

For a start the energy coming to the Earth from the Sun varies with time because solar activity itself varies over an eleven-year cycle. Not only that, but there are a number of other Sun cycles with different durations that have less of an effect. There is controversy over how much we know of the Sun's cycles in terms of their numbers and energy because we haven't had the equipment to measure these until relatively recently. There is also dispute as to how much they all actually affect global temperatures.

The position and orientation of the Earth with respect to the Sun will also affect how much energy the Earth gets and also how it subsequently distributes this energy across its surface and down into the depths of its oceans. Position and orientation change from one year to the next.

The Earth's tilt with respect to the Sun also changes in at least one cycle with a 46,000-year period. Its angle of precession also varies. To help explain precession, think of a spinning top. As well as spinning round its axis, this axis itself also spins around a central point. This cycle of precession takes around 26,000 years.

The orbit of the Earth is also changing in terms of its eccentricity, which is the amount it departs from a perfect circle. There are several cyclical components to this, each with different cycle lengths with one as long as 413,000 years.

However, this is not the end of the story as there are additional massive confounds arising from the way this heat is redistributed by the Earth. The part of the Earth's atmosphere that is head-on and nearest to the Sun will get the most heat. This heat will, over time, redistribute to the parts that are furthest away. Redistribution is at first through atmospheric convection which is itself highly complex, as exemplified by the inaccuracies of weather forecasting, but it also will be distributed to the much more difficult to heat up oceans and land. Redistribution of heat to the deepest parts of the oceans and the ice caps can take thousands of years and is affected by the circulation path of the currents around the world, such as the Atlantic conveyor, whose efficacy of redistribution is itself affected by temperature.

On top of this, the solubility of carbon dioxide in the ocean is also affected by its temperature. Additionally, when some of the heat is transferred to the ice caps the effects of their melting will affect the Earth's albedo which affects the amounts of heat the Earth can reflect back into space.

Complexity is mounting on complexity and mathematically nasty feedback loops are developing to make all our equations non-linear. In other words, unique solutions to these climate equations are pipe dreams. Solutions can only be produced if additional assumptions are made.

But it gets worse. So far, the changes in the source of heat from the Sun have been more or less regular, though the effects are complex. But there are other less regular contributions.

For example, there is also the matter of heat from the Earth itself, the centre of which is very hot. Indeed, the inner core of the Earth is thought to have a similar 5000 degrees Centigrade temperature to the surface of the Sun. The movement of huge tectonic plates that make up the crust of the earth, and the source of most of its earthquakes, can affect water circulation and thus how this heat from deep down reaches the surface. More acutely, the release of ash into the atmosphere from volcanoes all contribute to temperature change. Ash, in particular, reflects heat back into space and so can cause cooling. Very large eruptions may be responsible for climate changes large enough to cause mass extinctions.

Mount Pinotubo in the Philippines erupted in 1881 and sent vast amounts of dust and 20 million tonnes of sulphur dioxide into the atmosphere. This alone is estimated to have reduced the amount of radiation from the sun reaching the earth by about 2% and caused a drop in global temperature by 0.25 degrees Centigrade for two years.

You should also bear in mind that the world-wide redistribution of heat from whatever the source takes many years, so again the Earth's volcanic history has to be taken into account in any calculations.

Normal biological processes also can cause cataclysmic changes in climate, changes which dwarf even those that the most extreme climate change believers would lay at the door of mankind. The evolution of photosynthetic plants is thought to have caused massive glaciation around 2 billion years ago. Blooms of ferns in what is now the Arctic Ocean also caused huge temperature changes nearly fifty million years ago. Changes in the agricultural practises of mankind over the last few hundred years are another confound to this picture.

There is even some evidence that global climate may be related to the Earth's magnetic field strength and that too varies.

The Sun's output may also change over time and not just in the cyclic ways we have mentioned. We have no way of knowing one way or the other as satellite-based thermometry has only been measuring it for 50 years or so.

With so many confounding factors it is difficult to estimate the magnitude of the individual effect of each one. After all, it is not like a physics laboratory where one can reductively try to eliminate all but one effect that can then be studied exclusively.

Even more complexity intrudes when we consider other non-linear mechanisms which affect carbon dioxide in the atmosphere. These take the form of various feedback mechanisms we know about and perhaps many more that we don't. Of the known mechanisms these all produce positive feedback (in other words they make heating worse) with the exception of only one which produces a negative feedback. The latter arises from carbon dioxide fertilisation. The more carbon dioxide in the atmosphere, the more plant growth and the more carbon dioxide is broken down by the plants to release oxygen

Positive feedback mechanisms that make the situation worse include the so-called plankton multiplier. Hotter weather seems to produce less carbon absorbing activity in the oceans though nobody knows why. Higher temps also cause higher microbe respiration and so higher carbon dioxide emissions, and also a reduction in the growth of forests that could hoover up atmospheric carbon dioxide.

Finally, higher temperatures cause release of methane from swamps, especially in Siberia, and melting wetlands. Methane is another greenhouse gas.

If all this wasn't bad enough, new factors which haven't been taken into account in our climate change models are being found all the time.

A recent one is that the different parts of the energy spectrum put out by the Sun don't vary in the same way, though previously it had been expected that they would [6]. Researchers found a four to six-fold greater decline in UV than predicted based on what was happening to visible light when the eleven-year solar cycle was declining. This is important because ultraviolet dissociates atmospheric molecules and provides a major source of heating for the middle atmosphere, whilst it is the visible and infrared reaches that heats the lower atmosphere and earth's surface.

This all has major implications for the accuracy of our climate models. As the authors say: "Currently there is insufficient observational evidence to... fully characterise other solar cycles, but our findings raise the possibility that the effects of solar variability on temperature throughout the atmosphere may be contrary to current expectations."

With all these factors contributing and interacting with each other, climate modelling has to be done on computers. In fact, it needs the very best computers; itself a bad sign as it means so many variables are involved that many assumptions have to be made to do any computation at all.

Many of the parameters that go into climate modelling are differential;

in other words, they change with time based on the effects of other variables that in turn also change with time. Computationally it's like chasing one's tail. There are also uncertainties in these models regarding: sources and sinks of greenhouse gases, clouds which strongly affect the magnitude of climate change, oceans which influence timing and patterns of climate change, behaviour of the polar ice sheets which impact on sea levels. And those are the ones we know about. Who knows what other feedback mechanisms come into play?

With such a plethora of mutually interacting non-linear confounds one has to wonder what use climate prediction models are. So instead of computer models, let's consider what we actually do know of how the climate has actually changed in the past.

The last ice age ended only 20,000 years ago. Indeed, before the recent vogue for seeing evidence of global warming even in the face of colder winters, a similar cold period in the 1950s led to speculation that a new ice age was on the way.

More recently, climatologists really started to get concerned when the eight hottest summers for over a hundred years all happened in the 1980s. On the face of it that does sound bad and it led to the present emphasis on developing theories and models of climate.

But was it really so bad?

There have always been short term variations in our weather. There was, for example, a medieval 'warm period' which lasted from 1100AD to 1300AD, allowing vines to be grown as far north as Yorkshire in the UK. But there was also a 'Little Ice Age' between 1400-1850 AD in which the Thames frequently froze. None of this was due to man-made release of carbon dioxide as this was pretty minimal and constant up to the 1800s. We know too little about climate to understand why these events occurred, but they would have sent any Middle Age climatologists into a lather. Theories of impending ice ages or catastrophic warming would have been produced and all would have been wrong.

There are many other examples. From 1860 to 1995 the average temperature of land and sea increased from -0.4o C to 0.25o C, which might be held to corroborate the global warming idea, but it should be noted that temperatures fell between 1945 and 1950 and between 1960 and 1970.

Despite all the present anguish about global warming there is evidence that the average temperature in central England was higher in 1250 AD (10.5o C) than it is today (9.25o C).

The panic about global warming followed some very hot summers in the latter part of the last century that sent the climatologists in desperate search for a culprit. They didn't have to look far. Atmospheric carbon dioxide (as measured on a mountain in Hawaii) has increased in terms of parts per million by about 22% over the last fifty years. It has increased by 30% since the advent of the Industrial Revolution in the late 1700s when man began to burn fossil fuels in earnest.

The International Panel on Climate Change (IPCC), a UN intergovernmental body, uses climate modelling to predict a 2.5o C rise in temperature over the coming century [7]. Such a change would be faster than anything seen in the last 10,000 years.

Again, this sounds scary and indeed many climatologists will trot out this phrase for maximum effect. However, it needs to be pointed out that temperatures over that the last 10,000 years have been far more stable than at any similar time period in the last 150,000 years at least. Indeed, our present period of stability is arguably what allowed our modern civilisation to develop in the first place. However, it is set against a background of some very scary temperature changes that occurred entirely naturally.

Ice cores from 2.5km deep at the Vostok Station in Antarctica have been used to measure temperatures (for the Antarctic anyway) for the last 200,000 years, though you should bear in mind that this technique is subject to some of the previously discussed errors.

These samples indicate that the last ice age lasted from 120,000 to 20,000 years ago. Global temperatures have changed by 12o C over 160,000 years from -10 to +2o C. Any temperature changes made by man so far are trivial compared to this. In fact, during the last 100,000 years, periods of less than a hundred years have been found where temperatures changed by up to 6o C, in other words more than double what the IPCC predicts for this century. And remember that no fossil fuels were being burned by our few bedraggled ancestors at this time.

There is even evidence of a 7o C change in only 50 years about 11,000 years ago.

Set against the background of these huge naturally occurring temperature changes, how can we be sure our present measly 1o C rise since man started burning fossil fuels are due to us and not due to Mother Nature just going about her business?

Even worse, the implicit assumption of more carbon dioxide causing higher temperatures may be wrong. Greenhouse gas emissions over the last 15 years have continued to soar and dire predictions about mankind's future in the rapidly heating world have become clamorous. Multi-billion-dollar industries have sprung up to meet this terrible threat. The only trouble is that in this last 15 years, a period in which fully a quarter of all man-made carbon dioxide has been added to the atmosphere, air temperatures at the earth's surface have remained pretty much constant [8]. None of the twenty or so main climate models predicted this. All predicted big temperature increases.

Temperatures have risen by about 1o C over a time period of one hundred years, but now they have flattened out despite the accelerating use of fossil fuels.

This means that all major climate models, at the time of writing, have proved wrong. There may be many reasons for this (see earlier for a list of all the sources of error). Perhaps the rise by about 0.5o C during the 1980s and 90s which caused all this furore may simply have been an anomaly. Perhaps carbon emissions may not be as closely coupled to temperature rises as has been assumed.

Alternatively, there be other unknown mechanisms at play or perhaps we have under-estimated or over-estimated the ones we think we know about.

Clouds may be another culprit for helping us stave off global warming. In the climate models, clouds are thought to make things worse as they absorb more heat from the sun, but some scientists are now questioning this and wondering if they may have a net cooling effect.

The oceans are another possibility for where this supposed extra heat has gone, but again this is in conflict with the measurements. Like the atmosphere, the oceans have also flattened out in terms of their temperature over the last ten years and not risen as the climate models have expected.

Perhaps in desperation, some climatologists are now suggesting that the top 700 metres of the ocean where the temperature measurements are made are staying stable, but that somehow it is the lower depths of the oceans that are heating up.

The fact is that climate does change with time and far more radically than anything man could manage by burning fossil fuels. We may be seeing the start of big changes as the last 10,000 years of a relatively stable climate has been the exception rather than the rule. This may be happening with or without the burning of fossil fuels. There are certainly arguments for reducing our use of fossil fuels, not least because there is only a finite amount of them and because they produce a great deal of pollution. Man's burning of these fuels may indeed contribute to temperature rise but the truth is that nobody knows by how much because of the complexity of the myriad factors underlying the way our climate naturally changes over time.

14

Risk in Science: Juggling with Chainsaws

Science is like an edged tool, with which men play, like children, and cut their own fingers.

Sir Arthur Eddington, astronomer and physicist

Our scientific power has outrun our spiritual power. We have guided missiles and misguided men.

Martin Luther King

Some fool in a laboratory may blow up the universe unawares.
Ernest Rutherford, physicist, joking about the energy in the atom

By now I hope this book has shown that, whether you take the sceptical or heretical views of science, science at present is a long way from achieving its aim of understanding how the universe works. That in itself has an unfortunate consequence. Like Captain Ahab in Herman Melville's book Moby Dick, scientists in their search for the truth may be pursuing something that may be both unattainable and deadly.

And by deadly we mean potentially at the extinction level for mankind and perhaps for all life on Earth. There are unquestionably vast energies in the cosmos. We see that in our own Sun; at a distance of 150 million kilometres we can still feel its warmth on our skins. We see mysterious bursts of high energies that sleet down on the planet from space in the form of cosmic rays of unknown origin. Meanwhile down here on Earth we've even made hundreds of our own little suns and are always investigating new means of mass destruction.

By tinkering with physics and biology are we like children walking through a minefield? Could we accidentally unleash these huge energies, or produce some new biological organism that might wipe us out?

We've actually been ambling through this minefield for quite a while now with scientists at each stage rationalising away the possibility of catastrophe. So far, they have been right, though usually for the wrong reasons, but will our luck hold forever?

In this chapter I will review some of the risks we have taken and show how scientists have argued, often erroneously, that these risks were minimal. Usually they didn't even see the dangers coming at all. Even where the risk is anticipated it is usually dismissed by recourse to existing theory, even though the experiments themselves are often designed to verify some aspect of that very theory. On such circular arguments does the fate of mankind often depend.

And there's a big irony here. Scientists less hidebound by orthodoxy are often looking to find discrepancies with the existing theory. They are hoping for unexpected effects and the history of science is littered with examples of where one thing was looked for and quite another thing found. That the effect we may unexpectedly discover may wipe us out as a species is never factored into considerations of risk. This is because estimates of risk can only be based on what we think we know.

We're going to start by discussing some instances in physics where potentially catastrophic risks were argued away or were unanticipated and some of the disasters that have subsequently occurred. Later on, we'll look at just how risky the biosciences can be.

In physics, the best and most obvious example of an effect that proved to be unexpectedly harmful is that of radioactivity.

The discovery of radiation

Following the chance discovery of x-rays by Rontgen in 1896, Pierre and Marie Curie and others sought to determine their source. Ordering a hundred kilograms of uranium ore, the Curies physically processed the material themselves. After isolating the radioactive element radium from the ore, Pierre actually strapped a sample to his arm for hours at a time to see what effect, if any, it had on body tissues.

Their laboratory was full of what some called 'fairy lights' in the form of glowing and often radioactive minerals, but otherwise the lab was rather basic. A visiting scientist described their laboratory as a cross between 'a stable and a potato cellar'.

Working away, the Curies didn't use masks or gloves or lead shielding. Why should they? Why should a mineral be harmful if you didn't swallow it and so allow it to react chemically with the insides of your body? They were scientists after all and it was not like they were dealing with black magic.

Except, in a sense, they were doing just that because they were investigating something inexplicable in terms of the science of the time. Certainly, nobody had anticipated the harmful biological effects of this new form of radiation.

Marie Curie, who routinely carried radioactive samples around in her pockets, died of radiation induced anaplastic anaemia and also suffered from cataracts that were likely caused by radiation.

Her daughter also died of radiation related leukaemia and her second husband, Frederic Joliot, died of liver disease almost certainly contracted from working with polonium which was also radioactive. Her first husband, Pierre, as mentioned earlier, was killed when his skull was crushed by the wheels of a horse drawn vehicle otherwise it is quite possible he too would have succumbed to radiation induced sickness. Many of their colleagues in their own laboratory and in other laboratories around Europe and then later in America also died of radiation induced illnesses.

Even today Marie Curie's notebooks are considered too radioactive to handle without protection.

But these effects, lethal though they were, took time to manifest themselves so for a while taking radium salts as a tonic was quite the thing. Spas opened that purported, fortunately usually fraudulently, to provide exposure of patients to the curative effects of radioactive waters [1].

An extreme example is that of Eben McBurney Byers, an industrialist who, following an arm injury, consumed vast quantities of a patent medicine called Radithor over several years. This contained high concentrations of radium salts. Radiation induced bone damage was so extreme it caused holes to appear in his skull and the loss of his jaw. He is buried in Pittsburgh in a lead-lined coffin.

Many luminous watch dial painters, usually women, died or became seriously disfigured because their practice was to wet with their tongues the brushes they used to paint the luminous dials. In doing this they transferred radium-based impurities into their mouths. Cancers of the jaw and throat became common in these workers.

The sometimes quite casual use of radiation continued into the 1950s and 60s. I even remember as a child that a chain of shoe shops in the UK offered an X-ray of your feet. These machines were called Pedoscopes. Exposing children to radiation just to see if a shoe fits: probably not a good idea.

Nowadays exposure to radiation is rigorously controlled. This isn't because of what happened to the Curies but rather because of what happened to survivors of the next example of risky behaviour by scientists.

The Atomic Bomb

The story of the development of the atomic bomb is rather more gnarly than the simplistic accounts one normally reads. It also involves several examples of risk taking on a grand scale that are often left out of historical accounts.

Atomic physicists managed to do what Isaac Newton and other alchemists had only dreamed of: transmuting one element into another. Rather than using Newton's arcane and totally futile alchemical methods, Ernest Rutherford did it by bombarding nitrogen with alpha particles (helium nuclei), thus changing it into other elements, namely hydrogen and oxygen.

Despite this success Rutherford never really believed, even up until his death in 1937, that mankind could ever make use of this new world of atomic physics. He certainly never suspected that his work was paving the way, barely more than a quarter of a century later, for incredible weapons of mass destruction.

However, even then not everyone shared Rutherford's complacency. Nobel prize-winner Walter Nernst in 1921, unnerved by Rutherford's work, went so far as to say: 'We may say that we are living on an island of gun cotton'.

Few were so prescient. Luckily, most physicists at the time were ignoring results (twas ever thus) that didn't conform with their existing theories and so missed evidence of the chain reaction that would make the atomic bomb possible [2]. Otherwise, with so many of the top physicists of the day being German, it is likely that Hitler would have wound up with the first atomic bomb. Hilarity would no doubt have ensued.

The fuse that would ignite this 'gun cotton' was the neutron. For many years scientists were aware of the existence of the neutron but didn't believe some of the results they were getting. They knew from Rutherford that atoms could be split by heavy, higher energy charged particles like alpha rays but didn't accept that neutrons, which were much smaller and without charge, could do it too. In fact, atoms were being split by neutrons in laboratories in Zurich, Berlin, Paris, Cambridge and Rome without the scientists accepting what was happening. The scientists refused to believe what their instruments were telling them because it didn't fit existing theory.

When certain nuclei broke down under bombardment by energetic particles these released more neutrons which in turn could initiate other reactions. This rapidly increasing chain reaction precipitated the release of vast amounts of energy very quickly.

The idea that atomic disintegration could produce enough energy to create an atom bomb came not from theory but from an unexpected experimental discovery. This was made by Otto Hahn and Fritz Strassman in 1938. They had been experimenting with uranium to produce the lighter element barium. When the uranium disintegrated to make this breakdown product, a great deal of energy was unexpectedly released in the process.

The breakdown of uranium nuclei in particular was gaining attention and the process was given the title of nuclear fission. The word 'fission', incidentally, was coined originally to describe the way that bacteria multiplied. Despite the clear evidence for nuclear fission, and the awareness of the neutron cascade process even as late as 1939, there was still severe scepticism. Niels Bohr went so far as to produce fifteen reasons why such nuclear energy would never be exploited, and even Einstein at that stage did not believe that nuclear based power production was possible.

Leo Szilard, doing his own experiments that demonstrated a profligacy of emitted neutrons, was perhaps the first to get an inkling of the practicality of using a manufactured chain reaction to produce large amounts of energy. Sufficiently unsettled, he tried to get colleagues such as Enrico Fermi to keep their work secret rather than disseminate it freely.

Scientists of the Allied Powers, many of them German and Jewish and deeply alarmed about the fate of the world if Hitler came to dominance, began to consider whether a new kind of bomb could be produced using the energy released by nuclear fission. Many were driven by a fear that Nazi Germany might develop such a bomb first and hasten Hitler's military defeat of the Allies. Several distinguished scientists, including Einstein himself, convinced the American government to devote huge sums of money to developing an atomic fission-based bomb. This was a colossal gamble bearing in mind the drain on resources of waging even a conventional war.

The Manhattan Project was born and would cost over $25 billion in today's money and employ over 100,000 people at its height. Its goal was to achieve a brand-new form of explosive many orders of magnitude greater than achievable with conventional explosives. In this it was a triumph, though a triumph of engineering much more than of theory.

In essence, a fission bombs works by bringing together two lumps of enriched uranium-235. If brought sufficiently close, enough neutrons hit enough other uranium atoms to produce at least a self-sustaining chain reaction. If the two masses could be brought even closer then you get an accelerating release of energy.

Unfortunately, the theory wasn't up to predicting exactly the circumstances under which this would occur, depending as it did on a number of factors such as the speed of the uranium masses, scattering angles and ranges of neutrons. As ever, when the complexity of reality intrudes sufficiently to undermine our theories, then supplementation by experiment is necessary.

Otto Frish, who had discovered fission in the first place and had already nearly been caught in the effects of an uncontrolled chain reaction, was in charge of the group trying to better establish the critical point at which a chain reaction was initiated. In the process Harry Dagnian was the first of the Los Alamos physicists to sustain lethal radiation exposure from a bench top chain reaction. He lived only another 24 days.

Louis Slotin was a member of his team at Los Alamos charged with refining these measurements and he did this by physically bringing two masses of enriched uranium together. In doing so he helped establish the critical mass required for the first atomic bomb.

Slotin knew what he was risking and called it 'twisting the dragon's tail'. Eventually, after the war in 1946, the screwdriver he used to keep the masses apart slipped and the room was immediately filled with a bluish light. Instead of trying to get away, he tore the masses apart with his bare hands, dooming himself but saving the lives of others in the lab. Calmly, he then insisted that everyone stay where they were whilst he drew a diagram on a blackboard of their relative positions. This would allow them later to estimate the level of dose received.

Nine days later he was dead.

At least Slotin knew the risks he was running. Further limitations of the theory became evident when the first test blast explosion was made in the Alamogordo desert in an area called the Jornada del Muerto (Death Tract) near the village of Oscuro (Dark). The explosion proved to be much more powerful than anticipated by theory. Unfortunately, we can't tell by how much because much of the measuring equipment was unexpectedly destroyed. This equipment had been placed at distances that theory had erroneously indicated to be safe.

Perhaps the best description of the immediate effects comes from, of all people, the military man General Farrell: "*The whole country was lighted by a searing light with an intensity many times that of the midday sun... Thirty seconds after the explosion came, first, the air blast pressing hard against the people and things, to be followed almost immediately by the strong sustained awesome roar which warned of doomsday and made us feel that we puny things were blasphemous to dare tamper with the forces heretofore reserved to the Almighty.*"

Meanwhile Oppenheimer, the scientific leader of the programme and 'Father of the Atomic Bomb', was thinking of a line from the Bhagavad Gita: "*I am become Death, the shatterer of worlds*".

The scientists and engineers under Robert Oppenheimer succeeded in creating two other such devices. These were dropped on Hiroshima and Nagasaki killing over 100,000 people outright and dooming many more to radiation induced deaths in the years to come.

The Japanese quickly surrendered, though whether it was because of the bombs or the entry of the Russians into Manchuria has been subject to debate ever since.

Despite the capitulation of the Japanese, and the fearsome effects of the atomic bombs, the United States and the Soviet Union went on the design bigger and bigger bombs, in the end moving away from nuclear fission and towards nuclear fusion. Fusion of hydrogen nuclei is the same thermonuclear reaction that is thought to produce the Sun's energy.

It was found that a fission bomb was needed to trigger this much more powerful thermonuclear fusion reaction. In the US this new 'hydrogen' bomb would be called 'Super'.

So much more powerful than anything before would Super be that scientists began to worry that: 'An irresistible global chain reaction might be released by the Super, which would transform the entire planet in a short time into a flaming and dying star'. Ouch!

They were worrying because the Earth's atmosphere contains a lot of hydrogen and might become part of the chain reaction caused by the bomb. Some people referred to the possibility of 'setting the atmosphere on fire' but this would be a nuclear rather than a chemical 'fire' and so would be incomparably more devastating than even that description would suggest.

Despite the possibility of causing mankind's extinction, the group led by Edward Teller assigned only two physicists, Emil Konopinsky and Cloyd Marvin Jr, to look into this possibility.

Using existing theory, rickety and speculative and often just plain wrong as it was at the time, they came to the conclusion that it wouldn't be a problem.

Not everyone was convinced, however, and the matter was referred to a third, more celebrated scientist called Gregory Breit to look into this possible 'global chain reaction'.

Breit was well aware that many famous scientists before the war had used their knowledge of theory to show the impossibility of large forces ever being released from the atom. That demonstrated the provisional nature of any existing theory: you depended on it at your peril. This awareness must have put Breit under unbelievable pressure because all he had to assess whether such a catastrophic eventuality might occur was existing theory.

In the end, however, he concluded that based on the limited knowledge of the time (the Standard Model which would supposedly explain everything was still thirty years in the future) that such a thermonuclear reaction would not kill all life on Earth.

Luckily, he was right.

But he might well not have been. At the time many measurements had been made on nuclear processes but using equipment that could not possibly have measured many of the particles now believed to occur in nature. Could such particles, if released during these reactions, have caused other even more potent reactions or even just speeded up the chain reaction? Might, like the original atom bomb, the explosive yield have been much greater than predicted and what effect would that have had on these more esoteric reactions?

On the face of it, might it have been worth putting more than three men to work on considering this problem? Might it not have been worth waiting for further experiment to extend the theory? Might it not have been better to spend more time looking for any unexpected forces and particles produced by these reactions?

Apparently, the answer to all these questions was 'no', not with the Soviets breathing down the Americans' necks in the Cold War race to develop fusion bombs. The more pressure put on people, the greater the risks that are inevitably taken.

There were problems with theory again when it came to calculating whether a hydrogen bomb would work at all. The calculations were fed into ENIAC which was the state of the art computer at the time. The answer indicated that such a bomb would never be a realistic possibility. This led Teller, seen by many as the show runner of the H-Bomb just as Oppenheimer had been for the A-bomb, to declare that all the work done on it had been based on 'nothing but fantasies'.

However, new theory was developed by Stan Ulam that took a different but more complex tack on the fusion process. Even more calculations were required and these were done on the more powerful successor to ENIAC. It was called MANIAC.

You couldn't make this up.

The thermonuclear device, called 'Ivy Mike', was tested on the Eniwetok atoll in the Marshall Islands. Ironically this was near the site of where the Great White Whale Moby Dick sank the Pequod in Herman Melville's novel of that name.

Having learnt something from the previous higher than expected explosive yield from the first A-bomb, a safety margin of ten times was used to estimate how close people could be to the epicentre of the blast. Just as well, as the yield was much greater than expected from theory, producing a fireball 3km in diameter, a crater 50m deep and debris falling 48km from the blast.

But the safety margin was enough. It wouldn't be in a later test, which brings us to our next example.

The Castle Bravo H-Bomb

In the early days of the Cold War, when it came to thermonuclear weapons, the US and Russia had gotten into what could crudely be described as a big swinging dick contest. Each was trying to produce a bigger and bigger hydrogen bomb. This got so absurd that bombs weighing 50 tonnes or more were tested. These were utterly impractical as they were far too massive ever to be carried by planes or missiles.

And all this was going on in the days before underground testing so the bombs were being set off above Siberia and the Pacific, tainting the atmosphere and ourselves with radiation to this day.

In 1954 the US wanted to test a new three stage thermonuclear device. This had a fission bomb trigger like the bomb dropped on Hiroshima and this would set off a fusion bomb which in turn would set off a lithium-based fission third stage.

Theory expected a yield of around 5 megatons that would be over 300 times the explosive power of the Hiroshima atomic bomb that had instantly sent 80,000 Japanese to eternity. Bikini Atoll in the Marshall Islands was to be the lucky place where it was to be tested.

The Castle Bravo device, weighing in at a modest 10.7 tons, was placed on an island called Enyu on one side of the atoll.

Despite concerns on the day about the wind direction and the effect this would have on where radioactive fallout might land, the test went ahead.

Pushing the fire button on the device that was expected by theory to yield 5 megatons in fact yielded a 15-megaton blast, producing a seven-kilometre diameter fireball and leaving a two kilometres diameter crater. The mushroom cloud deposited significant levels of fallout over 18,000 square kilometres of the Pacific.

The firing team in their World War II bunker, positioned at what had been considered a safe distance of 24 kilometres away from ground zero, barely survived. Equipment to measure the blast was vaporised.

Fallout in the form of 'white flaky dust' rained downed for hours on Lucky Dragon 5, a Japanese tuna fishing boat around 40 km from the blast. One brave crewman even had a lick of what they later named the 'death ash' and described it as being 'gritty but without any real taste'. Aikichi Kuboyama, the radio operator, would die within the year from acute radiation syndrome. He is regarded in Japan as the 'First Martyr of the H-bomb'. The other crew members survived but had to be treated for beta burns, nausea and bleeding from the gums.

It wasn't just the Lucky Dragon 5 crew who were unlucky. Around a hundred other fishing boats were in the unexpectedly large fallout zone as were hundreds of Marshall Islanders. Whole islands had to be evacuated and some are kept under restricted access even now over seventy years later. In particular, islanders on the Utrik and Rongelap atolls were affected by the 160-kilometre-long fallout plume.

Even worse, the fallout appeared in rain over Japan and in winds over Australia, the United States and Europe, and entered the food chain many thousands of kilometres from the event.

All of what we've discussed so far happened in the past but now let's talk about some of the potentially far worse risks that are being taken by physicists today.

The Large Hadron Collider (LHC)

The LHC is a 27km circumference particle collider buried as deep as 175 metres below the border between France and Switzerland. It is designed to accelerate beams of protons or lead nuclei to within a fraction of the speed of light and then collide them.

Many people were worried about what would happen when the LHC was turned on for the first time and started colliding beams of particles at higher energies than had ever achieved before. Single particles would supposedly be given the energies of cruising battleships. Temperatures 100,000 times hotter than at the centre of the sun would be produced.

Concerns were raised that in the process black holes might be created, sucking the Earth into them and destroying all life. Indeed, creation of such microscopic black holes would have confirmed certain recent theories on modified gravity.

There's always an upside.

But black holes weren't the only potential sources of extinction level events. Others included the appearance of exotic creatures such as strangelets and Bose-novas and vacuum bubbles.

In response to this, CERN set up the LHC Safety Assessment Group [3, 4, 5]. Sounds innocuous but imagine what it was like for the members of such a group. After all, here was a $9 billion enterprise on which literally thousands of your physics colleagues from around the world were gambling their whole careers. Saying the whole thing was unsafe and should be halted would be a courageous thing to do.

On the other hand, if you got it wrong, all life on Earth might be extinguished.

It may well be that the members of this committee took the high road and all heroically rose above personal considerations to deliver a reasoned and impartial verdict. The trouble is that there is an inconsistency at the base of their reasoning which is summed up nicely by one supporter of the LHC [3]: '…for the questions of black holes, physicists built on existing scientific theories and data to evaluate the risks, and thereby determined there was no worrisome threat'.

There's obviously a circular argument at work here. The LHC was built to test whether existing theories were true or not, but these very theories were being used to 'establish' the safety of the LHC. And as we saw in Chapter 3 these theories are based on all sorts of questionable renormalisations (fudges by any other name), strange new dimensions, multiple universe and even entirely new forces. If physicists were so sure of the safety of the LHC then they would presumably be sure of the validity of underlying theory. There would therefore be no need to build the LHC to test aspects of the theory in the first place. Not, at least, when the fate of mankind might hang in the balance.

And this is all happening in a universe where vast but unexplained bursts of energy periodically blast through space and scour whole star systems. Perhaps each was the result a physicist hitting a switch, or whatever the alien equivalent of that is, on some even bigger collider somewhere.

Involving as it does the possible extinction of all life on Earth, this is about as important a point as there can be so let's look in a little more detail at the arguments for LHC safety.

The safety group didn't consider it out of the question that such black holes might be created. They also took into account that, in classical relativistic physics, once created these black holes would persist and grow as they pulled in neighbouring matter and so would have plenty of time to destroy mankind and every other living thing on Earth.

However, the safety group's get-out-of-jail free card was provided by Stephen Hawking whose theory indicated that small black holes could decay before they could do any such harm. They would do all this decaying by the emission of something called Hawking radiation.

All well and good, though the theory also included the idea that Black Holes heat up as they radiate and this rather flies in the face of everything else in the universe which radiates as it cools down. However, perhaps the more significant fly in this soothing ointment is the fact that Hawking radiation has never been detected.

So, if black holes were created, the Safety Group was gambling the fate of all mankind on Hawking radiation whose existence has never been verified. This was too much even for the committee, so they went a step further and asked what happened if Hawking Radiation didn't exist.

Their next argument is a little more indirect. They believed that some cosmic rays in space have the energies required to create black holes if they collide with other objects. So why haven't black holes formed before on the cosmic ray bombarded Earth?

It's a good question but it does make certain assumptions, one of which is that it is simply the energy that can produce black holes, though as we've never created a black hole or seen one being created then how can we even be sure of this? It also assumes that a cosmic ray striking atoms in the high atmosphere of the Earth is the same as identical beams of particles hitting each other in a carefully controlled laboratory experiment. One problem is that we know little of the original nature of cosmic rays. What we see of them are the showers of particles that are the breakdown products of collisions of cosmic rays high in the atmosphere. All we have are estimates of the energy of the original cosmic rays and this is being equated with the energy of the accelerated protons. Is it really just a matter of the (estimated) energies of the two being the same or is it possible that things might be a little more complicated than that? Just like everything else in the universe in fact.

Is this simplistic analogy, even in conjunction with the unproved existence of Hawking radiation, worth gambling all life on Earth on? What do we get in return? Some evidence to support an existing theory that history shows us will inevitably be superseded at some point by another?

The Heretic isn't worried by the idea of Black Holes forming because he doesn't believe they exist in the first place. As far as he is concerned they depend on the mathematical concept of infinity which he believes has no correspondence in reality. He is still worried however because there may well be dangers out there that in our ignorance we just do not anticipate. A hundred years ago, who knew that bringing two lumps of enriched uranium together could cause a chain reaction and a blast big enough to send over eighty thousand Japanese civilians to Kingdom Come? Or Madam Curie treating radium so casually, or radiologists working in the early days of x-rays, waving their fingers in front of x-ray tubes to show the bones moving in their fingers. Nobody anticipated the fatal effects of this radiation until they started to get ill and die.

Perhaps there are forces out there that we may accidentally unleash. Perhaps, as we keep on building more energetic accelerators, we are like someone on a fragile raft drifting downstream, oblivious to Niagara Falls ahead.

Physics itself provides other exotic examples of such potentially catastrophic hazards and these too might have come into existence when the LHC was turned on. The first one we'll deal with is strangelets.

Though perhaps more likely to be produced in heavy ion colliders than the LHC, these theory-predicted particles contain equal numbers of up, down and 'strange' quarks that might become stable rather than decaying away as quarks should normally so. When aggregated, such objects could become 'quark stars' or 'strange stars'. It has been suggested that because the strangelet's stability increases with size it might change any normal matter it came into contact with into strange matter. This cascading process would therefore convert all other matter, including you and me, into strange matter.

So, what would happen if you, me and all the rest of the world was converted? Difficult to say but if, for example, the net quark charge is negative rather than positive, then nuclei would attract rather than repel and the structure of every atom in our bodies would be catastrophically undermined.

No problem there then!

Despite strangelets never having been observed, and so their properties never having actually been established, the LHC Safety Committee found 'no basis for any conceivable threat'. That this conclusion may not be soundly based is nicely illustrated by a sentence in their report's introduction regarding the LHC: 'Many new and exciting phenomena are expected to occur as a result of these very high energy collisions. It is hoped that some of them will be unpredicted and will point to new directions in our understanding of the structure of matter. At the same time, it is legitimate to wonder whether any of these new phenomena may be potentially dangerous.'

If these phenomena aren't predicted, and therefore their properties unknown, how can anyone judge them safe based on analysis of existing theory that did not predict them in the first place?

Similar reasoning is used to dismiss the danger posed by our next potential hazard: Magnetic Monopoles. Again, it has been suggested such particles might be created, the effect of which would be to promote something called protonic decay, breaking down nuclei and releasing large amounts of energy as well as causing transmutation of the atoms. Who knows, perhaps all the atoms in your body would turn to platinum or gold. You'd be dead but you'd certainly make a good-looking corpse.

The argument goes that, even if this happens, it would have an insignificant effect. This is because each event supposedly causes the monopole to move. Safety factors are calculated on the basis of how many such interactions there would be before the monopole was shoved off the Earth and safely out into space. Enough, the equations indicated, that the monopoles would be kicked into space too quickly to have a significant effect, though of course this is again based on a number of theoretical assumptions, not least theoretical estimates of interaction cross section for which no experimental data exists.

However, in a later update [5] to the report it says: 'Nevertheless, if the magnetic monopoles were light enough to appear at the LHC, cosmic rays striking the Earth's atmosphere would already be making them, and the

Earth would very effectively stop and trap them. The continued existence
of the Earth and other astronomical bodies therefore rules out dangerous
proton-eating magnetic monopoles light enough to be produced at the LHC.'
Interestingly this issue of 'trapping' was not mentioned in the original report
which suggested the monopoles would be rapidly shunted into space.

This rather suggests to our Heretic that they making it up as they go along.

Bad though the possible destruction of mankind might be, it's as nothing
compared to our next hazard. The physics throws up the possibility of the
LHC destroying the whole universe via something called a Vacuum Bubble.

This is what the updated LHC Safety Report says about this: 'There have
been speculations that the Universe is not in its most stable configuration,
and that perturbations caused by the LHC could tip it into a more stable state,
called a vacuum bubble, in which we could not exist. If the LHC could do
this, then so could cosmic-ray collisions. Since such vacuum bubbles have not
been produced anywhere in the visible Universe, they will not be made by the
LHC.'

The main thrust against all these exotic but lethal effects occurring is that
cosmic rays containing even higher energies than the LHC can produce have
been bombarding cosmic objects such as the Earth, Moon, Sun and stars for
billions of years. Surely, we would have seen such catastrophic effects elsewhere in the universe if any of these dangers were real?

On the other hand, we know very little of the nature of original cosmic
rays themselves and even if we did, can we really compare these with what
may well be the much simpler particles we accelerate to high speeds near
Lake Geneva?

Incidentally, the above is not a comprehensive list of all the possible
catastrophic effects of running the LHC. In fact, in the update of the report,
more potential hazards were included than in the original document, such as
a 'runaway fusion reaction in the LHC carbon beam dump' and the possible
creation of a Bose Nova. If these hadn't been considered from the start and
have just been thought of, what other hazards are still out there but remain
completely unanticipated?

Destroying all life on Earth or even the universe itself may have disturbed
some readers so we're going to chill out a little now and consider the lesser
risks that come from the field of biology. These might lead to the deaths of
just a few billion people.

Archiving the smallpox virus

The 'spotted pimple' disease as it was known in Roman times can, even in its minor form, kill around 1% of sufferers. Its major form kills around 65%. It was dubbed 'smallpox' in England to differentiate it from the Great Pox (syphilis). Blindness and severe scarring can be consequences for people who manage to survive the major variety.

The disease is thought to have emerged in Egypt around 10,000 years ago and the first actual clinical evidence of it was found in the mummified remains of the pharaoh Rameses the Fifth. It went on to kill perhaps 400,000 a year in England alone in the 18th century, with mortality rates amongst children well over 50%.

The British even tried to use it as a bio-weapon in the 18th Century by giving contaminated bedding to indigenous peoples in North America and Australia. Beware Brits bearing gifts.

In the 20th Century it may have killed as many as half a billion in total. Vaccination in one form or another has been around for perhaps 3000 years but has been used in earnest only in the last two hundred years as we saw in an earlier chapter. Intense campaigns of inoculation were undertaken across the world. In 1979 the World Health Organisation certified that the disease had been eradicated. This was a major achievement; rinderpest, a disease of cattle and other ungulates, is the only other infectious disease eradicated by man. The last incidence of the minor type of smallpox was in a hospital cook in Somalia, the major version in a two-year-old girl in Bangladesh.

The eradication was a major triumph for practical as opposed to theoretical science. It brought about the end of a nasty disease that would never kill again. As a result, people are no longer inoculated against it and so millions, maybe even billions, might die if the disease was unleashed again. Good job the disease was wiped out.

The only problem is that at least one person has died since the disease was 'eradicated'. The person died not in the Horn of Africa, or in the Indian subcontinent, but in the large modern city of Birmingham in the United Kingdom. A medical photographer named Janet Parker died from the disease in 1978 from samples kept in a research laboratory. The head of the laboratory subsequently committed suicide.

Rather than destroying all samples of smallpox, many laboratories in the world had kept them. The death of Ms. Parker prompted a consolidation of all samples (or so we were erroneously told, as we will see later) into only two sites: the Centres for Disease Control and Prevention in the US, and at the State Research Centre of Virology and Biotechnology in Russia. Some scientists claim that the samples may be useful in producing future vaccines. Against what?

Though the WHO recommended destruction of all remaining samples as long ago as 1986, this has still not been done.

Thus, a scourge of mankind that has killed many billions, and could kill billions more, is still being kept alive in the safe hands of scientists. What could possibly go wrong?

We'll get a glimpse of the answer to that question in the next two sections, though the death of Mrs. Parker suggests that any faith we may have in scientists controlling such hazards may be naive.

Biological warfare

This is also called germ warfare because of its use of bacteria, viruses and even fungi to kill or incapacitate humans or destroy their crops. It's been around for at least 2500 years since the Assyrians laced enemy wells with a fungal agent that incapacitated rather than killed their enemies. In the 14th century the Mongols, those ancient charmers, catapulted dead plague victims over the walls of cities they were besieging.

The big problem with bio-weapons is that they are just so damned difficult to control. Unleashing a virus on the enemy could easily backfire if it spread to your own troops or civilian population. In recognition of this, the Biological Weapons Convention in 1972 outlawed the mass production, stockpiling and use of biological weapons. By 2011, 165 nations had agreed to this treaty.

So, does that mean we are OK?

Not quite, because it still allows nations to research into defences against biological weapons and permits them to create new viruses to test these defences. That's all fine as far as the Convention is concerned as long as they don't mass produce them.

Research into bio-weapons is fertile ground. For a start, nature has been plentiful in supplying bacteria and viruses that can be tailored to mass killing. Here are some of the beauties that have been studied and, in some cases, weaponised:

Ebola virus (characterised by explosive bleeding from just about every-where, followed by death in nearly 50% of those who catch it)

Marburg virus (symptoms too long to list but all are nasty. Has caused at least one death in the laboratory in the USSR where it was being weaponised)

Rift Valley fever virus (bleeding from the eyes and mouth, black vomit, liver damage)

Various bacteria to produce cholera, pneumonia, septicaemia and, of course, the bubonic/pneumonic plague.

Anthrax offers a particularly exciting potential for mass killing. Spread by spores, this allows better control of where it is distributed. Not only that but it produces 90% fatality rates in those infected but untreated. It is not considered transmittable from one person to another and you can, in theory, protect your own troops with antibiotics. There are, however, a number of assumptions in this theory and one wonders if it would still apply if genetic alterations were made to the anthrax organism as seems to be common practice. Due to the complexity of even this relatively simple organism, the results could well be unpredictable.

But bio-weapons don't just need to be new diseases or existing diseases altered to be more virulent or transmittable. One can take a less direct approach by, for example, producing agents which nullify the effects of any vaccine or antibiotics administered for an existing disease within the target population. Or you can take a disease which doesn't affect man and tinker with it until it does. Even more ambitiously, work is presently underway to change existing diseases to be undetectable by the usual diagnostic means. Fiendish!

Modifying diseases in any of these ways requires an in-depth and comprehensive knowledge of all the biological processes of the disease organism itself and of the organism that it infects. Our buzz-killing Heretic would argue that such a depth of knowledge will never be achievable so bio-weapons produced will never be as controllable and their effects never as predictable as we hope. The Sceptic might agree that that is presently the case but believes that someday, for good or bad, we will be able to understand these issues and produce effective, controllable bio-weapons.

Even if never deployed, the accidental release of these pathogens is possible even in highly controlled research environments. People die, for example Janet Parker with smallpox and Nikolai Ustinove with Marburg in the USSR. Is it too cynical to imagine that, sooner or later, an example of this 'defensive' work is going to escape from the laboratory into the environment and a new plague will sweep the world?

After all, everyone makes mistakes.

And such escapes have already occurred. One such example, which it is believed to have caused around a hundred deaths, occurred in 1979 when weaponised anthrax escaped from a military research facility in Sverdlovsk,

1500 km east of Moscow. Supposedly caused by a blocked filter being re-moved but not replaced, escaping spores drifted directly across the street in the northerly wind and into an adjacent ceramics factory. Had the wind been in the other direction, the spores would have blown back into the city with far greater consequences. The incident was labelled a 'biological Chernobyl' though the USSR repeatedly denied the story and put the deaths down to tainted meat [6]. It has been reported that the type of anthrax, coded as 836, in fact had come from a previous accidental release in the city of Kirov in 1953. Rats trapped nearby had been found to carry a particularly virulent strain of the disease and the strain in the Sverdlovsk leak had been subse-quently isolated from that.

That all happened over thirty years ago so surely things are a lot better now when it comes to the safe handling and control of such agents? I'm afraid not.

Here's what the Guardian newspaper had to say in an article on 19th July 2014 called Handle with care. Staff working at the Centres for Disease Con-trol and Prevention (CDC) in Atlanta followed the wrong procedure to deac-tivate the bio-terrorism agent Bacillus anthracis. While the spores were still lethal they were sent to another lab that wasn't designed to handle live spores. By luck, nobody got the disease.

In another incident, a supposedly harmless strain of bird flu was sent by a CDC lab to the US Department of Agriculture. In fact, it was contaminated by a highly lethal strain, one that has killed hundreds of people.

Again, nobody died but these incidents so alarmed the head of the CDC that he closed the labs working on anthrax and influenza. He called the be-haviour of his staff 'totally unacceptable'.

Six phials of supposedly destroyed smallpox were discovered in an un-guarded FDA lab in Maryland (as mentioned earlier, only two centres were supposed to have samples of smallpox). Other samples found with the small-pox were Dengue Fever, influenza and Q fever. The head of the NIH thought what had happened was 'unacceptable'.

As the Guardian said: 'To the outside world, the most trusted keepers of lethal germs had shown themselves to be dangerously incompetent'.

Bearing in mind that research into 'defence' against bio-weapons is highly secret, there are likely to have been many more such incidents that have never been reported.

Many scientists believe the risks of handling these things are too great, even for the highly controlled labs of the CDC and NIH, never mind other less regulated setups in Asia and Russia. Even Europe has its problems: the UK's Pirbright Lab was most likely the source of the 2007 foot-and-mouth outbreak.

These more cautious scientists say fewer labs should have access to this material (in the US alone there are a staggering 1,500), and high-risk experiments should never be attempted even under 'secure' conditions. For example, in 2011 the virologist Ron Foucher announced he had made bird flu far more transmissible by making it spreadable by coughing and sneezing. These 'gain-of-function' studies are supposed to help scientists better understand diseases but if their bio-security fails they could cause rather than prevent a pandemic.

As Sir Richard Roberts, a British Nobel Prize winner, says on the issue of bio-security: 'How can you trust anybody? Humans are human. People make mistakes.' And 'If I suggested that you try to make the most virulent and dangerous virus that we can imagine, something that could kill a quarter of the world's population if it got out, does that seem a sensible thing to do? That strikes me as being absolutely ridiculous.'

Probably the most exciting development on the horizon if you're a bio-weapons scientist, but less so if you're a potential target, is the use of targeted bio-weapons that might exploit the genetic differences between the races. They're called 'ethnic bullets'. As well as killing the race you disapprove off, other options include targeting their reproductive systems so they would give birth to deformed children. Or you might, more kindly, simply reduce their fertility so that over time their race would just vanish from the face of the Earth without making too much of a fuss.

Bearing in mind that the genetic variations between individuals are often more than between races, this sort of weaponry is still beyond us. Of course, the Heretic would point out that our understanding of the complexity of the whole genetic process (DNA, metabolomes, transcriptomes, epigenetics etc etc) is so sparse that such a bio-weapon could never really be so targetable. Indeed, our incomplete understanding might lead to any such agent being more likely to cause the end of the entire human race.

HELA and the contamination of samples worldwide

Laboratories that deal with biological samples are scrupulous in their sterile procedures. I mean, science is all about control, surely?

Dead wrong on both counts.

The reason we know this is because of an African American woman in Maryland in the United States who died many years ago. Despite being long dead, some say she is the first ever immortal.

Henrietta Lacks died in 1951 from an extremely aggressive form of cervical cancer. The cells of her cancer were harvested and used as samples for studies evaluating possible new cancer therapies in the laboratory. The unique thing about her cancer cells was that they continued to robustly reproduce even after they had been taken out of her body. Many other attempts to create an immortal cell line, defined as one that lasts for more than a few cell divisions, had failed with maximum survival times being only a few days.

So robust were Henrietta Lacks' (HeLa) cells that they were bred and sent to other laboratories around the world. Some have pointed out that assessing therapies by using this uniquely hardy and aggressive cancer for testing cancer therapies was perhaps setting the bar a little too high. Indeed, it may be that many useful cancer therapies were abandoned in their early stages of development because of HeLa's ferocity.

The robustness of Henrietta's cells has, however, been found useful in the testing and production of new vaccines, such as Jonas Salk's polio vaccine, as the cells were found to be easily infected by the disease. Their longevity is the key here as they stay alive long enough for the virus or bacteria to kill them. HeLa has also been useful in studies of AIDs and the effects of radiation. HeLa cells were the first human cells to be successfully cloned and the first cells to be cultured in space.

So ubiquitous are Henrietta's cells in laboratories around the world that it has been estimated that if all the samples of her cells were brought back together they (she?) would now weigh over 20 tonnes and getting heavier all the time. Not bad for the cells of someone who died over sixty years ago. It has also been estimated that by 2009, over 60,000 scientific papers had been published of experiments using her cells, with around 300 new ones continuing to appear every month.

HeLa and Henrietta Lacks herself make a fascinating story [8] but they also lay bare a dark side to biological sciences to do with the issue of contamination. This is because HeLa started to turn up everywhere.

It is vital in any biochemical research for samples to be kept free of all bacterial and viral contaminants. In the 50s and 60s a huge amount of research money had been spent on developing cell cultures from tissues and organs. The point was to study their development and how they responded to drug interventions or environmental changes. A cell culture industry had developed and conferences had sprung up, not the least of which was the Decennial Review Conferences on Cell Tissue and Organ Culture.

At the second of these meetings in 1966 a man called Stanley Gartner gave a paper that was described by a future President of the American Type Culture Collection as follows: 'He showed up at that meeting with no background or anything else in cell culture and proceeded to drop a turd in the punch bowl.'

The Cell Culture Collection had been a national collaboration to build up a library of cell cultures that had been tested for, and found uncontaminated by, any form of bacterial or viral contamination. What Gartler basically described doing in his paper was taking 18 samples of different types of cell line from this collection and finding a specific contaminant in each. The contaminant was HeLa. As it was another human cell, rather than an alien bacteria or virus, none of the contamination checks had been designed to look for it. This was because human cells had never been considered aggressive enough at such low levels to essentially take over tissue cultures. But HeLa cells, whilst human, are aggressive to the point where even a single cell, wafting in the breeze or piggybacking on a dust grain, and landing in a nutrient solution in which a cell culture sample was being grown, becomes all-conquering. It displaces the normal cell culture, like a cuckoo hatching in another bird's nest.

But the really shocking thing Gartler showed was that it wasn't even that 18 cell lines had traces of the same contaminant. Instead, all eighteen of these supposedly different cell lines were in fact exactly the same. All of the samples had become Henrietta Lacks' cancerous cells.

As well as undermining a decade of work by teams of scientists, it also put the lie to a phenomenon that many researchers had been reporting and believing in, namely the 'spontaneous transformation' of normal cells to cancerous ones.

Of course, it was all just because Henrietta Lacks' cells had taken them over.

And it wasn't only in the field of cancer that this came as a bombshell. People had been growing cultures from the cells of specific organs to investigate their normal biochemical functions. Papers were being published and textbooks written about these functions for organs like the liver. In fact, what they were writing about were the abnormal functions of the diseased cells from the cervix of long dead Henrietta Lacks.

As invariably happens with all heretics, it took quite a while for Gartler's work to be believed as many scientists were in a state of denial. They continued to blindly use their 'contaminated' samples for years. Eventually, though, Gartler's revelations did kill off the whole field of research on 'spontaneous transformations' and no further reports of such a thing have been heard of since.

As late as 1972, and on the other side of the world in Russia, HeLa cells were still taking over sample cultures and leading to erroneous conclusions. In this case it involved the supposed discovery of a 'cancer virus' which again turned out to be HeLa cells.

Walter Nelson-Rees made something of a pariah of himself when he started to produce 'HeLa hit lists' where he checked cell cultures used by other scientists in their experiments. He published those that had been erroneously identified and that were in fact HeLa cells. In doing so he sometimes invalidated many years of work by whole teams of scientists. According to a book by Michael Gold 'A Conspiracy of Cells' Nelson Rees stopped doing this in 1981 due to budget cuts from the National Cancer Institute but also because of the personal difficulties his work had caused with other scientists.

Without this watchdog it is possible that HeLa cells are still causing their mischief today. Certainly cross-contamination, and the misidentification that comes with it, is still considered a problem according to the International Cell Line Authentication Committee. Studies are therefore still appearing in the literature which may be entirely fallacious as researchers rarely check the purity of their cells.

Even today scientists are perhaps in a form of denial about how innately rocky and vulnerable our understanding is of the real world. They put the whole sorry saga down to the robustness of the cells, rather than to problems with human understanding and diligence and the levels of separation and sterility used in laboratories. Today HeLa is dismissed as a 'laboratory weed'.

Some weed!

So much for biologists exercising control over dangerous organisms in their laboratories! Now let's look at a dangerous idea that springs from evolutionary theory.

Eugenics

Sir Francis Galton was a remarkable man. Perhaps most renowned for his creation of the idea of statistical correlation, he also devised the first weather map. Not only that but he was the cousin of Charles Darwin.

Oh yes...and he was also the person who coined the term 'eugenics'.

Gripped by his cousin's Origins of the Species, Sir Francis began to consider whether man might take hold of the reigns of his own evolution rather than being at the mercy of nature. He devoted much of the rest of his life to the measurement of man and indeed invented the science of psychometrics that is supposed to measure peoples' mental abilities. He also looked at characteristics such as fingerprints for which he produced a classification scheme.

Convinced that intellectual ability was primarily hereditary (nature rather than nurture) he advocated that couples with high status should be granted money as an incentive to produce more children. As eugenics goes, this was as tepid as it gets.

His ideas caught on and in the early 20th century many countries had adopted, usually covertly, eugenics measures such as segregation (both racial and the segregation of the mentally ill from the rest of the population) and marriage restrictions. More overt methods included forced abortion, sterilisation (many countries did this for the mentally handicapped including even moral stalwarts like Sweden and Canada), euthanasia and then, later, full-on genocide by the Nazis. Proponents of eugenics included Marie Stopes, George Bernard Shaw, H.G. Wells and Winston Churchill.

Though the Nazis were extremely successful in giving eugenics a bad name, it has been said that our modern attempts at exploiting genetics are simply eugenics by the back door. Whereas the mapping of the first genome of a man cost billions, the cost has dropped to become almost trivial. The genome of the foetus can be mapped and, if considered unsatisfactory by whatever societal or parental standards prevail at the time, can be aborted. This book has tried to show how little we actually know about how we get from DNA to the full human being; it is quite possible that in the years to come many millions of perfectly viable foetuses will be aborted based on poorly understood information.

There are other less direct but perhaps far more significant consequences. By breeding for specific traits such as height, we may also reduce the inherent variability that has allowed man to adapt over the millennia and thus climb to the top of the food chain. If we artificially breed out our own variability then we may even be paving the way for being supplanted at the top of this by some more versatile species if there was some significant environmental change to which we could no longer adapt.

The most desirable trait to breed for in the modern world is that of intelligence. We really haven't a clue what genes, if any, code for intelligence but over the coming years we can be certain that some scientists will claim they do. For such a complex trait, even the Sceptic may be led to the conclusion that any attempts to select for this, by whatever eugenic means, are as likely to produce the opposite of the wished-for effect.

Designer babies are those who's DNA has been mapped before implantation in the womb using in vitro fertilisation. It may be possible to modify the technique of gene therapy, presently used to try to treat certain diseases, to actually change a baby's DNA in vivo. This would make the future ripe with unintended consequences.

For example, testing for and eliminating some traits in order to eliminate defects may be a worthwhile aim, but unless we know the full story these attempts may make the baby prone to other diseases or cause chronic defects of their own. An obvious example, and many of these effects will be far less obvious, is the case of sickle cell disease that untreated can cause a forty-year reduction in life expectancy. The thing is, this also provides immunity against malaria and protection against tuberculosis and cholera. Cystic fibrosis also provides similar protections.

All of this comes down to how we categorise such 'defects'. One man's defect is another species' prize. For example, Down's Syndrome and autism certainly increase neuro-diversity and may only look like defects from the 'normo-centric' point of view at this moment in time.

In the next two sections we're going to look at how far down some very dark alleys our limited understanding of biology can take us.

Thalidomide

The unwanted effects of drugs have been known for many years [7]. Again, if you need convincing, then look at the long list of side effects in the literature that comes with any prescribed medication. Even something as apparently benign and useful as aspirin can cause fatal haemorrhages.

In the early part of the 20th Century an attempt was made to regulate drugs for the first time. This took the form of the Wiley Act in the US in 1906 and was a response to quack medicines that could contain anything from opium to dung. This Act allowed the Department of Agriculture's Bureau of Chemistry (now the Food and Drug Administration) to examine food and drugs. The main concerns were about the ingredients in drugs, additives and the adulteration of food. A later amendment to the Act looked at claims of how well drugs worked. However, whilst fraudulent claims would be considered illegal, claims based on sincerely held views, even if erroneous, were not made illegal. Such belief was, of course, a rather difficult thing to prove or disprove.

One case in 1937 is worth mentioning. The Massengill Company in Tennessee sold sulponamides (anti-bacterial agents) as a liquid rather than the more usual powder, liquefying the drug by the expedient of dissolving it in something very similar to antifreeze (diethylene glycol as opposed to antifreeze which is ethylene glycol). This additive, like antifreeze, was toxic. Nevertheless, no animal or human testing was even considered. 107 people died and the company's chemist, Harold Watkins, killed himself. Even so, the company itself denied liability on the grounds that they could not have 'foreseen the unlooked-for results'. The government of the day wanted to bring fire and fury down on the company but it turned out there was no law prohibiting putting a lethal substance into a medication. In the end the company was caught out on a technicality. The drug was being sold as an 'elixir' which by law meant it must contain alcohol, not diethylene glycol. The company was fined for mislabelling.

This was as outrageous then as it is now and it prompted the passing of the 1938 Food, Drug and Cosmetic Act which required that a drug's safety be demonstrated before it was marketed though not, as it happens, before it was used. Standards of proof were not high. Double blind randomised controlled trials hadn't even been conceived of at that point in time.

In the 1950s a company called Chemie Grunenthal in West Germany produced a drug that had no effect when tested on animals. It had previously been tried out by a company in Switzerland but had been abandoned as a benign but therapeutically useless drug. However, one of the chemists at the German company thought it had a similar molecular structure to barbiturates and so might give the company entry to that highly lucrative market. Neither of the companies who had investigated it had found a sedative effect but nonetheless that is how it was marketed. One can speculate that the company perhaps had more faith in chemical theory than experiment.

Before the drug could become available commercially, it was given to some doctors to try on their patients. Anecdotal reports from some of these doctors reported that it helped their patients sleep. This was music to the company's ears so, to get a license to sell the drug, they went back to their labs and managed to produce a test where mice given the drug appeared to move less. Hardly evidence of a sedative effect, but this was enough to get a license.

Released originally under the brand name Contergan, thalidomide was touted as a sedative but was later recommended to treat morning sickness as some reports seemed to show it was good in reducing vomiting.

The company now described it as 'the drug of choice' in treating morning sickness and 'the best sleeping pill in the world'.

In the end around 10,000 children suffered severe physical deformations due to their mothers taking the drug, with half of them dying in childhood.

A dogged drug reviewer called Frances Oldham Kelsey working for the United States FDA had turned down the application for a licence. She'd worked on the 'elixir' of sulphonamide scandal so had past form for sniffing out bad medicines. Her action no doubt prevented many babies being born deformed in the US but it was not enough to prevent this entirely. Though they couldn't sell it there, the company could simply give the drug to doctors in the US so children did suffer, though in much lower numbers than in countries where it had gone on sale.

The thalidomide scandal was enough to prompt more stringent testing of both safety and efficacy not only in the US but also in the rest of the world.

There is a twist to the tale of thalidomide and one which nicely illustrates the often arbitrary nature of drug development. As we've seen before, a drug may often be conceived from theory and then developed to treat one thing but is found fortuitously to help with another, even if it fails to help the condition it was developed for.

Erythema Nodosum Leprosum is a form of leprosy where clumping in the blood vessels, caused by the body's own antibodies fighting the invading germ, clog up the micro circulation. The vessels are unable to supply oxygen to tissues downstream from the blockage so the tissue dies. Such tissues include skin, nerves, joints, eyes and testicles. Skin drops off and ulcers appear both outside and inside the body.

The pain can be terrible to the point where morphine's numbing effects last only minutes rather than hours.

One such sufferer came to Jacob Sheskin in the Jerusalem Hospital. Running out of palliative options, Dr. Sheskin found an old bottle of thalidomide pills. He was aware of the original claims, however dubious, of their sedative effects. He knew of the effects on embryos and the nervous system but his patient was an old man whose peripheral nervous system had already been severely affected by the leprosy.

Not only did the man's pain radically reduce following administration of thalidomide but his wounds even began to heal. A placebo-controlled randomised trial that followed on other patients showed the effects were general to the condition.

Since then, thalidomide has been found efficacious in multiple myeloma (a form of cancer) where the drug's chemotherapeutic effect is being utilised, and also in Silk Road Disease which involves the small vessels and where thalidomide's immune system modifying effects come into useful play.

Thalidomide has also been found useful in eliminating the mouth ulcers that can come with HIV, again by reducing the immune response and so reducing inflammation.

None of this had been anticipated by theory when the drug was originally developed. What had been anticipated never materialised but, in the process, it brought tragedy to tens of thousands.

DDT

Dichlorodiphenyltrichlorothane is an insecticide that was first synthesised in 1874. Again, though it would eventually be found useful for unexpected purposes, it wasn't originally designed to be an insecticide. In fact, these properties were not discovered for another sixty years; Paul Muller being given the Nobel Prize in 1948 for doing this.

Readily adapted to powder or solvent or vapour delivery (even candles could be made to contain the stuff) it became the insecticide of the 1940s and 50s and was used to destroy the insect vectors of typhus, dengue fever and malaria in particular.

Huge quantities, nearly two million tonnes of the stuff, were released into the environment without any consideration of its ecological effect. Indeed, what happened next was the thing that kicked off the environmental movement in the 1960s in the first place.

What scientists had not taken into account was that birds eat insects. If you kill all the insects then the birds die too. Whilst insect populations can quickly be restored after the effects of the single application of DDT has worn off, birds will take a lot longer to re-establish a viable population. Birds are a major factor in keeping down the insect population. In other words, a single application of DDT can be highly effective in the short term but can cause greater long-term problems, which in turn require more and more applications of the stuff to keep the insects in check.

Another effect not anticipated at the time was that DDT resistant strains of insects would be produced. This meant that many of DDT's gains, such as the elimination of malaria in some parts of the world like India and South America, were lost as insect populations re-established themselves. These new populations, being DDT resistant and with the bird population reduced too, actually made the situation much worse.

Not only that, but it was suggested that DDT caused cancer of the breast, liver and pancreas. The substance was to be eventually banned from use under the Stockholm Convention. The aim was to eliminate, or at least restrict, the use of this and other persistent organic pollutants, though DDT's total elimination may take until 2020.

DDT was only one of a 'dirty dozen' of the worst compounds. It was found that DDT and its breakdown products were lipophilic (fat loving) and stayed in the body for many years. DDT in humans has been linked with diabetes and a reduction in reproductive capability by reducing both the quality of semen and the duration of lactation. There is also an increased chance of miscarriage, due mainly to its disruptive effect on hormones and the endocrine system. Evidence of toxicity to the nervous system of the human foetus has also been found, leading to developmental disorders and decreased cognitive skills.

DDT works, it is thought, by opening sodium channels in neurons and causing them to fire throughout the brain in something like a massive epileptic fit. After severe spasms the creature dies. Again, as an illustration that compounds rarely have single effects and not only on the target organism, it was found that the breakdown products of DDT caused eggshell thinning and therefore greater mortality of the embryos of some birds of prey.

Despite the salutary example of DDT, the intervention of science, usually for the best of reasons, continues to have unintended consequences. Treatments for malaria, for example, are thought to have been responsible for the recent and increasingly common appearance of the drug resistant and much more malignant version of malaria caused by the parasite Plasmodium falciparum.

We know so little but it doesn't stop us tampering with complex systems and situations. Inevitably there are unforeseen consequences.

Science gives us theories that purport to explain how the universe works. This breeds confidence in scientists who then go on to do things that carry certain risks. These risks are rationalised away on the basis of existing theory. Even if our Heretic is wrong in saying that all theory is actually erroneous, history shows us that most or perhaps all theories ultimately prove incorrect. Our perceptions and calculations of risk are therefore also likely to be erroneous. Science generally also assumes a high degree of control over experimental conditions and again this faith seems misplaced. While we may routinely underestimate risk, we also routinely overestimate our ability to control it.

And this is set against a universe brimming with devastating energies, dangerous physical entities and deadly organisms.

Where does this all leave us?

15

So Now What?

It is an illusion that something is known when we possess a mathematical for-mula for an event; it is only designated, described, nothing more.

Nietzsche [1]

People make the mistake of talking about 'natural laws'. There are no natural laws. There are simply temporary habits of nature.
Alfred North Whitehead, mathematician and philosopher

Then (in around twenty years time) nanotechnology will allow us to live forever. Ultimately nanobots will replace red blood cells and do their work thousands of times more effectively. Within 25 years we will be able to do an Olympic sprint for 15 minutes without taking a breath, or go scuba diving for four hours without oxygen.

Ray Kurtzweil [2]

'So now what?' is a good question. In this chapter we'll try to summarise where the Believer, Sceptic and Heretic think we go from here.

But first we need to be aware that there are problems appearing in conventional science, both in terms of its rapid fragmentation into smaller and smaller sub-specialities in order to try to explain an ever more complex universe, but also by its rapidly diminishing returns.

About 150 years ago the number of scientists, or their equivalent of the day, in the whole world could be counted in the hundreds. These were often religious men who had time on their hands such as Church of England parsons who were especially active in making observations of the natural world.

Nevertheless, despite their low numbers, the discoveries of these early scientists came thick and fast and were often made simultaneously by more than one of them. In contrast there are perhaps six million professional scientists in the world today and the tools at their disposal are incomparably more powerful than anything that was available then. Compare the slide-rules then with the super-computers of today; the small refractor and reflector telescopes of the 19th century with the Hubble Telescope, never mind the other instruments now available like radio telescopes that can scan the heavens in wavelengths far beyond visible light. Think of the resolution of the humble light microscopes then and electron microscopes now.

The question is, with perhaps a ten thousand-fold increase in the number of scientists, a million-fold increase in resolving power of many of our measuring devices and a trillion-fold increase in calculating capacity, why are major discoveries and developments appearing so slowly? The news that stars are accelerating away from each other is the only really profound discovery I can think of in the last half century. Not only that but fundamental mysteries still remain despite science's vast resources (some of these are briefly discussed in the Appendix).

The Believer and Sceptic would say that this slowdown in major discoveries is simply a matter of the low hanging fruit already having been picked and that we will just have to work harder and harder to discover new laws of nature or to further our understanding.

The Heretic would take a completely contrary view, as ever. He or she believes the idea of laws underpinning reality is a falsehood and as a result we need more and more scientists and more and more computers to produce greater and greater elaborations on our theories to make them fit inconvenient experimental data. We're forced to make our theories increasingly complicated and also to break science down into smaller and smaller sub-specialities, each with ever more disparate theories pertaining only to their speciality and not to others. More and more 'special case' laws are generated for each speciality because they just don't work if they are applied more broadly. The laws of the very small do not apply to the very large. Neither Maxwell's equations of electromagnetism in the energy domain, or the various laws of thermodynamics in the molecular domain, apply to each other or to the atomic or nuclear domains. The more laws and principles we have to generate to make our theories fit reality, the less likely there are any underlying laws in the first place.

And even that isn't working. Physics, the hardest of hard sciences, is finding itself having to resort to concepts that can never be experimentally verified such as multiple universes and a plethora of new spatial dimensions. For an explanation of the physics of the very large and the very small, many physicists have quite literally taken the step from science to faith.

In this book criticism of science has been generally within its own terms and using its own methods. However, if science is beginning to come off the rails then perhaps the problem lies with its most basic tenets. Let's step back a bit and consider these. Science makes a number of baseline assumptions which are usually not explicitly acknowledged and for which there is no actual proof [3]. Among these are:

Everything is essentially mechanical in the sense that people are just atoms and molecules that interact via underlying laws. No mystery there, no room for the soul. The love you feel for your children? Just nuts and bolts. Heroic self-sacrifice to save others? Biochemical processes pure and simple.

All matter is unconscious. Even the human 'consciousness' is just a series of hard wired feedback loops in the brain. For example, perhaps you believe in homeopathy but then you find it doesn't work for something you suffer from. The brain monitors this, recognises it as a problem and then (hopefully) significantly modifies or abandons this belief.
Supposedly, these feedback loops are essentially what constitutes your mind.

The total amount of matter and energy in the universe never changes, though this assumption is beginning to fray at the edges as modern cosmologists have to resort to the idea of a vast cosmic burp, or vacuum fluctuation, producing the whole universe out of nothing.

There are laws of nature and they do not change with time and place and they are based on mathematics.

There is no purpose to the universe. Cosmic development and the process of evolution are blind. They are not designed to go anywhere.

(There are scientists who are also Christians but they seem to be able to cope with this last assumption which rather conflicts with their religious beliefs by compartmentalising their lives. For example, to do evolutionary science that will be published in a reputable journal, they have to pretend that the universe has no purpose even though they also believe it was made by God for a reason. If they research in neuroscience then they must pretend brain functions can be broken down into electrical or biochemical effects. If they want to get their work published in prestigious science journals like Nature then there is no room for any input from the soul, though that is the basis of their religious beliefs)

The Believer regards all these assumptions as gospel, so to speak, but the Sceptic may be a little more nuanced. More open minded than the Believer, and aware of some of the problems that modern science is having, he or she is perhaps willing to entertain the idea that one or more of these basic assumptions are incorrect. Nevertheless, he or she still believes that if an incorrect assumption was finally identified as such, then the scientific method would modify itself to continue with the same onward and upward scientific progress that the Believer assumes is the case anyway.

The Heretic would say that not only are all these assumptions as evidentially unfounded as a belief in God but would go further and point out that they are as meaningless as our scientific laws. Like them they are simply projections of our brain-limited way of thinking onto a universe which just is what it is.

That said, the Heretic is making one assumption himself, though it is one for which there is overwhelming evidence, namely that everything in the universe is unique. Science, by generalising everything, is ignoring what is perhaps the only real truth of the universe, possibly it's only underlying law.

As well as a possible law for uniqueness, the Heretic is open to the possibility of a second law governing complexity, namely that it increases with time. Evidence for this comes from the burgeoning intricacy of life on Earth as the eons have passed. Whether uniqueness and complexity are really 'laws' or just further examples of the universe just being what it is, the Heretic doesn't really mind much either way.

If uniqueness is the only real truth then it undermines all our mathematics that can never be anything more than an approximation, so there is good reason to believe that at least the fourth assumptions above is fundamentally wrong.

The Heretic accepts the universe has certain patterns or regularities of behaviour we can exploit, for now at least. Whether these regularities that scientists believe are governed by hard and fast laws or principles really stay the same with time is something nobody can know. Something else that nobody knows is whether these regularities are the same everywhere in the universe.

Whilst the Heretic may doubt all these assumptions, the Believer and his many scientist friends never challenge or even think about them at all. Instead they continue their wide-eyed dreaming of unlimited progress as illustrated by the work of people like Ray Kurtzweil (see quote above). Though these views are extreme, they are surprisingly well supported. Kurtzweil's concept of the technological singularity comes from the idea that even in down-and-dirty reality some things can really go to infinity and do so in mathematically predictable ways. In this case what is going to infinity is the growth in human knowledge and the speed of computation. Proponents consider that at the moment when computers become more intelligent than humans we will embark on a massively accelerating period of technological growth. Computers will then design and build even smarter computers who will in turn build even smarter ones. To infinity and beyond!

The event horizon of a black hole is sometimes used as an analogy to represent the point when we come second in intelligence to computers and the 'intelligence explosion' kicks in. Beyond this our poor limited minds will not even be able to understand what is happening because it will be the rapidly evolving computers that are doing everything for us. Kurtzweil and others predict we will cross this event horizon in the 2040s, give or take.

Many singularity proponents consider that these super intelligent machines will give the poor old humans a lift up the evolutionary ladder. Our handicapped minds will somehow be boosted by their technology. Our minds may even be downloaded onto silicon, our thoughts vastly accelerated and our memories made far more reliable and rapidly accessible.

Of course, even to the Sceptic, never mind the Heretic, this is all complete nonsense. With 100 billion neurones in each human brain and without any real clue as to how they are all integrated together to form the human mind, this whole idea slips far beyond extrapolation into something very akin to religious faith.

Nevertheless, Kurtzweil and his supporters like Peter Diamandis, founder of the X-Prize that encourages technological developers, went on to found something called the Singularity University in 2009 to 'educate, inspire and empower leaders to apply exponential technologies to address humanity's grand challenges'.

The university is concerned with familiarising students with advances in nanotech, artificial intelligence and neuroscience. The University makes a fetish of the exponential and points to Moore's Law as a prime example. This says that the number of transistors on integrated circuits will double approximately every two years and was coined by Gordon Moore, a co-founder of Intel. Moore's law is in a sense self -fulfilling as long-term planners in chip companies use it to map long term strategies for the funding of future chip development.

Even though Moore's Law is often cited by the singularity believers, Moore himself says that his prediction of exponential growth would not last indefinitely.

But Kurtzweil has many supporters. According to Wired [4] Larry Page, co-founder of Google, has appointed Kurtzweil as Google's Chief Engineer and is funding him to develop artificial intelligence, primarily with a view to allowing computers to actually understand human language.

Kurtzweil isn't the only famous scientist who paints a picture of the future that would warm the cockles of our Believer's heart. Craig Venter was one of the scientists who helped decode the human genome. He did this independently via his own private company and not as part of the internationally funded initiative. Venter, rather cheekily, used his own genome as the sample to be decoded. He has since gone on to invent new life in the sense that his team designed and then constructed a simple microbe. He believes that DNA is just computer code and so is readily subject to manipulation, allowing even greater improvements in things like crop yield and drought resistance, as well as the construction of synthetic animals that could be used for food or as biological factories in the production of drugs [5].

But, of course, the Heretic would say that this again suffers from the assumption that life is, at heart, made up from relatively simple processes. As we have seen, the work with the human genome project has not yielded the expected breakthroughs in medicine or the understanding of disease. It has simply scratched the surface and found an ocean of complexity dropping vertiginously away beneath. Without real understanding of such complexity, dabbling with the genome of any organism is likely to be ripe with unintended consequences.

Nevertheless, the Believers of this world have a future replete with weird new animals, interstellar colonisation, God like computers, and people augmented to the point where they aren't even human any more. The Sceptic can maybe envisage all this happening, though only far away in the future. The Sceptic's view is that our present theories are far from the truth, but that the truth really is out there.

However, the Sceptic is not blind to the possibility that we may need to ditch some of the basic assumptions and really start to think out of the box if we are to make many more truly profound scientific discoveries and developments.

Some sceptical scientists have been trying to think out of the box for a while now without actually straying far into heresy. For example, Teilhard de Chardin was a scientist in the fields of both geology and palaeontology in the early part of the last century. He was also a Catholic and in fact was trying to slip God back into the mechanistic world of science. Nevertheless, his ideas are worth discussing as they do suggest an alternative way of thinking that would still be essentially science based.

He proposed the concept of an Omega Point which was a form of hidden attractor working on all the supposed laws of physics, chemistry and biology. This, in effect, led them to produce more complexity and organisation and also led to the development of human consciousness. To him the universe was inevitably progressing to this Omega Point at which time it would have reached a stage of maximum organised complexity. This Omega Point attractor could perhaps be thought of as the equivalent of a company's 'mission statement', supposedly reflecting the universe's underlying ethos. That said, it would have to have infinitely more substance than any mission statement I've ever seen.

Of course, there is no actual evidence for hidden attractors or Omega points.

A more recent thinker out of the box is Rupert Sheldrake. He has written extensively on his belief in the existence of 'morphic' fields. These, for example, give form to an organism over and above any instructions encoded in its DNA. They may also dictate how proteins fold into only certain chemically active 3D shapes and also organise the crystalline forms that liquids can take when they cool. Once such a form has been first created then this supposedly affects the appropriate morphic field and means subsequently that an identical liquid, even if the sample is on the other side of the world, will find it easier to adopt this newly created form.

As evidence of this he cites cases of where liquids, once they take up new crystalline forms, seem to supplant the old forms to the point where these are no longer possible [3]. Conventional chemistry does, however, take a counter view. Crystals require a 'seed' to produce them. Once a new, more optimal crystalline form spontaneously appears, its seeds are then carried on the winds or by researchers themselves (their beards are supposed to harbour seeds of all kinds) and this is the way the new crystalline form is transmitted.

Sheldrake's ideas are interesting in that they do challenge some of the basic assumptions of science. For example, he suggests that even inanimate matter has a form of consciousness. He believes each atom undergoes experiences, for example at one point it may 'feel' the attraction of an adjacent atom. This history in turn affects its morphic field and affects the ways it will interact in future.

Sheldrake's ideas come with little hard evidence or, to be more precise, evidence which convinces most scientists. However, Sheldrake makes the point that research funding bodies rarely make funds available to really try to test these claims. Funding bodies tend to take the established view as the truth and generally fund only studies that will test extensions to the existing theory. They almost never fund tests that may undermine it. Why use limited funds to question what is already accepted when you can spend it on sexy new research?

Whether their views are right or wrong, it is good there are always doubters around like de Chardin and Sheldrake though the road they take is hard and underfunded. Without them the basic unproved assumptions of science will never be challenged and further profound discoveries may never be made.

Let's summarise the views of our three knockabout characters: the Believer, the Sceptic and the Heretic.

The Believer is happy with the basic assumptions of science. He doesn't rule out that there will be strange experimental findings to come but, rather than undermine theory, these will simply show that further elaborations and modification will be required. Diseases will be chipped away at until we are all completely healthy and one day maybe we will travel to the stars. Our minds will be downloaded onto computers and we will become immortal all-knowing Gods.

Oh yes, and there'll be jet packs for all.

The Sceptic says we actually know very little. Vast discoveries are therefore still out there to be made, some perhaps involving massively destructive energies. Perhaps the sudden energy spikes that occasionally appear from space are what happens when an alien race takes an incautious step in their research. He understands we may be wandering blindfold through such a minefield but thinks the risks are worth it for the sake of progress. He accepts the possibility that some of the basic assumptions of science may be wrong but he's not willing to give up on them yet. Even if one is wrong, for example that the laws of physics don't change with time, then a new physics could be created to take that into account.

He also believes that we will eventually be able to genetically modify ourselves in significant ways. One day our science driven technology will probably take man to the stars.

The Heretic, grumpy and cantankerous as ever, takes a glummer view though not perhaps without its own spiritual resonance. If theory is never right because it is the projection of our limited human ways of thinking onto a universe that does not follow them at all, then theory should be abandoned. Perhaps we should stick to simply looking for regularities or trends in the universe's behaviour in the hope we can exploit them using our engineering prowess. We should try to get the emphasis of research back to open minded observation and away from attempting to corroborate or disprove rigid theories. Not being shackled by theory in this way may actually be quite liberating.

However, as some examples from the history of science have shown in Chapter 14, there may be dangers out there that we might stumble across and perhaps even end our existence. If the universe does not abide by our man-made scientific laws then we're never going to see these coming as any predictions we make will be based on these fictional rules. At the very least the Heretic urges the scientific community to greater caution. Perhaps, indeed, there should be greater oversight by non-scientists: people whose own careers are not dependent on such research.

The Heretic believes we'll never download our minds onto computers and will rarely be able to significantly alter our genetic code without producing unfortunate consequences. There may be reasonably effective treatments for disease still to be discovered but they will generally be found by discovering naturally occurring active molecules and not by designing new ones based on theory.

And he doesn't believe we'll ever get to the stars (see the section on the 'Fermi Question' in the Appendix for the reason why).

In all this the Heretic is completely out of tune with the spirit of the times and will immediately be accused of being a defeatist. He or she seems to be saying that as a race we should hunker down and try to avoid trouble by stopping our scientific endeavours or at least exercising a great deal more care than we do at the moment. Just because such beliefs are out of tune with current 'can do' thinking does not mean they are untrue.

One trouble with the heretical view is that most experiment today is driven by theory. The theory may be wrong, may indeed always be wrong, but it is the impetus for experiments that would show up any of these supposed regularities in the first place. If we wish to continue to really improve ourselves technologically then some new impetus would have to be developed to direct experimentation. At the very least we need to free ourselves of the mindset that there are hard, invariable underlying laws governing everything.

And if we do continue to experiment then we need to take a very different approach. The Heretic suspects we should be looking to adapt the sort of qualitative research practises used in the Social Sciences. There, some attempt is made to factor in the effects of the experimenter who can never be an entirely disinterested and 'objective' observer. These attempts to explicitly acknowledge the cultural or social biases of the observer, and even to try to find ways to reduce their contribution, chimes in with the Heretic's belief that the way our brains are hard wired to think innately biases our view of the universe and our understanding of how it works.

The Heretic appears at first sight to be offering us a very confined place in the universe as he or she believes our most seemingly powerful tool, scientific theory, is at best rudimentary and perhaps even completely incorrect. At the very least, the Heretic is saying that we need to profoundly re-assess how we look at the universe.

But, on the other hand, who knows what profound insights may be revealed when we remove these blinkers?

Appendix: Some Fundamental Mysteries Still Not Explained by Science

Many years of effort by countless intelligent scientists have not even come close to convincingly answer a number of rather basic questions. Here are just a few of them. Taking the heretical view may not exactly solve these mysteries but perhaps it can provide some insights into some of them.

1) Why do we sleep?

Why we sleep is a colossal mystery. Why on earth should creatures like ourselves spend nearly a third of our lives unconscious of our environment? While we are asleep we are exquisitely vulnerable. That might not be such a big deal nowadays where we have lockable houses and the rule of law but it's much more of a big deal if you're yielding consciousness while lying out on the savannah or in a jungle somewhere. Predators often hunt at night to exploit this very vulnerability.

Not only does sleep increase our profiles lion-bait wise, but lack of sleep can kill you all by itself. For some reason sleep seems to be essential in all animals that have a brain. The only possible exceptions may be certain ocean dwellers such as sharks that have to keep moving so that water flows over their gills. Even then it's possible that sharks, like some marine mammals, are actually sleeping one brain hemisphere at a time.

Whilst we still don't know why we sleep there have been a number of hypotheses put forward.

Some say it is to reduce metabolic activity so that the body's various systems can repair themselves after a hard day's living. On the face of it that sounds reasonable but what about creatures that hibernate? When hibernating they don't move around or do things that they would need to recover from. The strange thing is that these creatures continue to have periods of sleep while they are hibernating. Why on earth does something that is hibernating need to sleep as well? Clearly there's something quite profound with the issue of sleeping.

Some say sleep is an evolutionary advantage to species like ours. It's a mechanism to stop us moving around in the dark so we don't hurt ourselves, or stumble into the path of a predator that hunts at night. The trouble is that this doesn't explain why cats sleep at all, as they can see both at night and during the day.

Also, as evolutionary mechanisms go, this seems a rather harmful one.

Such mechanisms usually arise to increase our chances of survival so we can breed. However, if you're totally deprived of sleep you will die. Even partially depriving someone of sleep increases their chances of developing all sorts of illnesses from Type 2 Diabetes to heart disease. This all makes sleep a very odd evolutionary mechanism.

It also doesn't make sense because it suggests creatures at the top of the predatory food chain should sleep the least as they have less to fear from the night. The trouble with this is that big predators like lions sleep the most of all the mammals.

Another theory has it that sleep is for the processing of memories acquired during the day. Certainly, if you are deprived of sleep then your ability to remember things is diminished. But organisms that don't need good memory to survive also go to sleep. Can it really be worth it in survival terms to render them unconscious and vulnerable for long lengths of time?

Some also say, and we're really reaching now, that sleep is an evolutionary hangover from when it was the animal's last line of defence from a predator. In other words, sleep is essentially 'playing dead'. Eight hours blind and unconscious seems quite a price to play for our prehistoric possum days.

Our present understanding of sleep has been described by the oft quoted line from William Dement who was co-founder of Stanford's Sleep Medicine Centre. After 50 years of researching into sleep, he came to the conclusion: "As far as I know, the only reason we need to sleep that is really, really solid is because we get sleepy' [1].

The need to sleep has other unfortunate consequences and not just because it makes us vulnerable while it's happening. Perhaps 20% of serious road accidents are associated with sleepiness; sleep disturbance is also much associated with depressive illnesses and human unhappiness. Some syndromes such as Fatal Familial Insomnia actually cause deaths in the sufferers.

Science generally seems to deal with the incredible mystery of sleep by ignoring it. Sleep research, in the words of John Winkelman, Medical Director of the Brigham and Women's Hospital Sleep Health Center "...just gets no respect".

Sleep is something that is clearly very profound and may well be intimately linked with whatever the mind is. Our Heretic would say that that is why we can never understand it in the same way we may never understand the mind itself. The mind can never, as it were, transcend the mind to get an overall perspective on the mind and its processes.

2) Has the universe a beginning or an end?

As we saw in Chapter 4, cosmologists are resorting to the concept of vacuum fluctuations, sort of cosmological burps from which whole the universe emerges, to explain where the universe came from. However, where did this bubbling vacuum come from in the first place?

It's like with God. If you believe God made the universe, then who made God?

Desperate cosmologists aside, most people recoil at the idea of the whole universe springing from nothing. Our Heretic would go further and say that there is no such thing as nothing in the first place; that it's just a consequence of our artificial mathematics. Despite our hang-up on the reality of the 'number' zero there is nowhere in the universe where we have found a region of 'nothing'. Even the space between galaxies has atoms and particles and photons travelling through it, not to mention the quite possibly entirely fictitious presence of Dark Energy and Dark Matter.

The Heretic would probably go even further and say that this whole business of beginning and ending is just a consequence of our linear brains. He would say that our beginning and ending are concepts that are simply a product of our limited life-spans where everyone and everything without exception comes into existence and then pops out of it. In days gone by, even the sun seemed to die every day at dusk and was born again at dawn.

Are we simply projecting the confines of our own existence onto the universe, just as we try to project our human mind notions of physical laws onto its workings? Is an absolute beginning and ending only of meaning to our time limited selves? Why does the universe have to come from somewhere? Maybe the Heretic is right and it just is.

3) The Fermi Paradox

With billions of galaxies, each containing billions of stars, the likelihood of life appearing somewhere other than just on Earth seems overwhelming. This has led many people over the years to ask the question: where are the aliens and why haven't they visited us?

The most famous person to ask this question was the physicist Enrico Fermi. This is why it is called the Fermi Paradox and is based on the Drake Equation which uses seven broad factors to estimate the number of alien civilisations that might exist. Factors include things like an estimate of the chances of a sun having planets that could sustain life. That factor was unknown in Fermi's day but nowadays it is thought that planetary systems around stars are common.

Fermi made a lot of guesses for the different factors in the Drake Equation and came to the conclusion that there would be enough other civilisations out there that Earth and all the other solar systems in the galaxy should be colonised by now. He estimated it should take at most 50 million years for an alien race to colonise the entire galaxy.

If you've never heard of the Drake Equation before your first reaction may be that this is all nonsense and, to a large degree, you would be right. The factors of the Drake equation (another example is the fraction of planets that can sustain life) are pure guesswork. But the point is that even if you give all seven of the factors in the equation very low probability values, this is set against the 400 billion odd stars in our galaxy. However low you make the factors, you still wind up with there being at least some high-tech alien civilisations out there.

So why aren't extraterrestrials here on Earth already? There are a number of answers to the Fermi Question, including:

That life is in fact unique on the Earth and that we are alone in space, though the Drake Equation would suggest this is highly unlikely

That intelligence is very rare and that we are one of the few species to develop it. Maybe, but it's difficult to understand why that would be the case

That intelligence aliens develop but, at some point, they become so technologically powerful that sooner or later they destroy themselves, as we ourselves came close to doing during the Cuban Missile Crisis of 1962. Having lived through that time myself I can see where that argument is coming from

There are countless better developed civilisations so we merit little interest. After all, would you bother to check out every ditch by the side of road for signs of life? Highly uncomplimentary but possible

There are countless more advanced civilisations but they all follow a kind of non-interference Star Trek-like Prime Directive when dealing, or rather not dealing, with newbies like the human race. Uncomplimentary and patronising

They may be here but are too advanced for us to detect their presence. Scary; let's not go there

Civilisations are out there, communicating all the time, but are using an advanced technology to do so. Being able to detect and interpret their signals would grant access to the Cosmic Club. So far, we've failed this entrance test

Civilisations are carried away by extinction events before they get advanced enough for interstellar travel. Such events might include nearby supernova or gamma ray bursts or large meteorite impact.

Our jolly Heretic can provide three other possible explanations:

The first one is that, like us, other civilisations do not understand how the universe works but also, like us, kid themselves that they do. They may therefore be unaware of extinction-level traps for the unwary. For example, occasionally we observe huge gamma ray bursts and other unexplained high energy phenomena coming from somewhere in the universe. Is each of these an alien civilisation making an unexpected but fatal discovery? Examples of some of the risks that our science has already taken are described in Chapter 14. Civilisations with the Believer's attitude that they are close to understanding the universe may be like babies playing with hand grenades. Sooner or later they manage to pull the pin out.

The second explanation is that if the civilisations manage not to kill themselves, they become aware of how deep their ignorance is and stop playing with the hand grenades. They settle for their lot and do not attempt to travel the vast distances between the stars.

The third explanation is that interstellar travel is simply not feasible. Again, the hubris our technology gives us makes us assume that, some day, interstellar travel will be possible. For example, many believe that faster than light travel will become available, possibly by the construction of wormholes. The Heretic would say the very idea of wormholes comes from a misplaced belief in the reality of the mathematical concept of infinity.

Alternatively, even if faster than light travel is never possible, as Einstein's Special Theory of Relativity indicates, many still believe mankind will one day build huge spacecraft holding large numbers of occupants that can still travel quite fast. Generations would live and die during these thousand-year voyages even to just the nearest stars. Others suggest the spacemen might be cryogenically frozen and then defrosted when they got to their destination. Or perhaps we would send out unmanned machines on our behalf that would eventually send back their findings like our interplanetary probes send back data from Mars.

But the trouble is, could we really make any type of complex technology that would work for even hundreds never mind thousands of years? The more complex the technology, the more likely it is to fail and often in ways that were never expected. You'd need someone to maintain it, which perhaps rules out the cryogenic option (if that was even possible in the first place). Would any group of maintenance engineers, cooped up for generations in such bizarre circumstances, last for a thousand years or more?

It's a humbling and disappointing thought that basic practical constraints on the reliability of our technologies and ourselves would stop us travelling to the stars, but it would explain why nobody else seems to be coming the other way.

4) Where are memories stored? In fact, how come you're still you even though virtually every cell in your body has been replaced since you were born?

This issue was mentioned in the chapter on the brain but let's look at it a little further. As Francis Crick, co-discoverer of the structure of DNA, said: 'Almost all the molecules in our bodies, with the exception of DNA, turn over in a matter of days, weeks or at the most months. How then is memory stored in the brain so that its trace is relatively immune to molecular turnover?'

Never mind just memory. If you don't even have the same atoms you did ten years ago then how come you're the same person you were then, give or take any 'learning experiences' an unkind Fate may have thrown you in the meantime

Crick was actually a determined mechanist in that he believed we are just made of biochemical nuts and bolts. In order to square this apparent circle, Crick came up with a complex scheme that would answer his question, though no sign of this has ever been found. It involved the body somehow carefully replacing one molecule in the brain at a time [2].

Massive head injuries can take out an entire brain hemisphere, whole lobes of the brain can be removed during surgery, large volumes of brain can be destroyed by cancer or stroke, yet all of these often have little or no effect on memory. This has led materialist scientists to come to the conclusion that memories aren't stored in one place in the brain but are somehow shared around it in a form of redundancy that can defeat such large-scale trauma. The hologram analogy is always dragged out at this point though there is no evidence to support its application to the brain. Holograms were originally produced from images on photographic plates. These plates can be broken up but each piece still retains information about the whole image, though at a lower resolution. It's an interesting analogy for memory but it doesn't tell us how that might actually be achieved in the brain.

We also have the issue of hydrocephalus where 90% of the brain can be replaced by fluid but the person still remains a normally functioning person [3]. We've also seen how 'simple' creatures with very, very little brain like termites can build complex structures and exhibit intricate behaviours.

These kinds of mysteries have led some people to at least begin to question whether memories (and perhaps even the mind, whatever that is) are stored in the brain at all. Indeed, one can even question as to what exactly the brain is for if we can live without so much of it.

Research into the workings of the central nervous system is one of the largest and best funded areas of science and has been for many years yet, somehow, we're still not even close to understanding these basic mysteries.

References and Further Reading

Books that are particularly worthy (or otherwise) of further reading are described within the sections for each Chapter below:

Introduction

[1] Paul Feyerabend. Against Method. Verso 1975

[2] A recent example of the Believer's idea that we more or less already know everything significant about the universe, can be found in 'The End of Science: Facing the Limits of Knowledge in the Twilight of the Scientific Age', by John Horgan, published by Broadway Books 1996. In the later British edition published by Abacus, the author tries to square the circle of the newly 'discovered' need for concepts such Dark Energy and Dark Matter that some would regard as compelling contrary evidence to his argument. His response included describing the new findings as 'trivial'. He also resorted to trying to redefine what he had meant by his use of the word 'fundamental' so that it didn't include this unanticipated new force that makes matter accelerate away from other matter, thus defying gravitational attraction. As we will see, the author is in the company of many distinguished scientists who have made the same mistake over the centuries of thinking the game of knowledge was almost over. The fate of these, if they lived long enough to see the mounting contradictory evidence, was one of fulsome retraction or, as in this case, rather unconvincing attempts at minimisation and redefinition. If you want a book that is in some ways a counterpoint to Science for Heretics then Horgan's book may be for you.

Chapter 1. Numbers shmumbers: why being 'bad with numbers' may actually be a good thing.

[1] Friedrich Nietzsche. Beyond Good and Evil. Penguin Classics 2003

[2] Anthony Beevor. Stalingrad. Penguin 2007

[3] Anne Rooney. The Story of Mathematics. Arcturus Publishing 2008

[4] David Berlinski: Infinite Ascent. Weidenfeld and Nicolson 2006

[5] Charles Seife. Zero The Biography of a Dangerous Idea. Souvenir Press 2000

[6] John D. Barrow. The Infinite Book. Jonathan Cape 2005

[7] Paolo Zellini. A Brief History of Infinity. Allen Lane 2004

[8] Edmund Husserl. Crisis of European Science and Transcendental Phenomenology Northwestern University Press 1970

[9] George Lakoff and Rafael Nunez. Where Mathematics Comes From. Basic Books 2000. Fascinating book that examines in exhaustive depth the idea that maths is entirely human brain based and not something innate to the workings of the universe.

[10] Eugene P. Wigner: Communications on Pure and Applied Mathematics 1960;13:1-14

Chapter 2 Science of the Very Large

[1] John Gribben. The Universe: A biography. Allen Lane 2007

[2] Lisa Randall. Knocking on Heaven's Door. Bodley Head, London 2011

[3] Paul Feyerbend. Against Method. Verso 1975

[4] Ian Angel and Dionysios Demetis. Science's First Mistake. Bloomsbury Academic 2010. This is a good, though rather academic treatment of some of the issues described in this book, particularly with regard to the problems of observation.

[5]
http://cosmictimes.gsfc.nasa.gov/teachers/guide/1919/guide/gravity_bends_starlight.html

Chapter 3. Science of the Very Small

[1] Lisa Randall. Knocking on Heaven's Door. Bodley Head, London 2011. This is a good general book on cosmology and, though taking a conventional view of the subject, is quite frank about the predictive limitations of present particle theories.

[2] John Gribben. The universe: a biography. Penguin 2008

[3] George Gamow. One, two, three…Infinity. Facts and Speculations of Science. Dover Publications Inc 1989

[4] Peter Woit. Not Even Wrong. Jonathan Cape 2006.
This is a book told from the point of view of a mathematician who believes in the reality of mathematics but takes offence at the credence and funding given to String Theory which he believes is a complete dead end.

[5] Feynman, Richard P. QED, The Strange Theory of Light and Matter, Penguin 1990, p. 128

[6] John Schwarz. History of Original Ideas and Basic Discoveries in Particle Physics Edited by Newman H., Ypsilantis T. Plenum Press 1966 page 698

[7] Paul Dirac. The Evolution of the Physicist's Picture of Nature. Scientific American May 1963, 45-53.

[8] A. Pais. Einstein and the quantum theory. Reviews of Modern Physics 51, 863-914 (1979), p. 907.

[9] Nancy Cartwright. How the Laws of Physics Lie. Clarendon Press 1983. This is a good book if you want to explore some of these ideas in more academic depth. It describes both the limitations of quantum concepts but also in some of the more general and supposedly prosaic laws of physics. She provides an interesting quote from particle physicist James Cushing:

When one looks at the succession of blatantly ad hoc moves made in QFT [quantum field theory] (negative-energy sea of electrons, discarding of infinite self energies and vacuum polarizations, local gauge invariance, forcing renormalization in gauge theories, spontaneous symmetry breaking, permanently confined quarks, color, just as examples) and of the picture which emerges of the 'vacuum' (aether?), as seething with particle-antiparticle pairs of every description and as responsible for breaking symmetries initially present, one can ask whether or not nature is seriously supposed to be like that?

Chapter 4. Black Holes and other Mythical Beasts

[1] Leo Tolstoy What Is Art and Essays on Art. Oxford University Press, 1930, trans. Aylmer Maude.

[2] Lisa Randall. Knocking on Heaven's Door. Bodley Head, London 2011

[3] Albert Einstein. Geometry and Experience, January 27, 1921

[4] Ray Kurzweil, The Singularity is Near, pp. 135–136. Penguin Group, 2005.

[5] Vinge, Vernor. originally in Vision-21: Interdisciplinary Science and Engineering in the Era of Cyberspace, G. A. Landis, ed., NASA Publication CP-10129, pp. 11-22, 1993)

[6] Stephen Pinker Tech Luminaries address singularity. Spectrum.ieee. org. Retrieved 2011-09-09.

Chapter 5 The Great Physics War and other skirmishes

[1] Samuel Butler. Notebooks Cape 1951

[2] Charles Seife. Zero The Biography of a Dangerous Idea. Souvenir Press 2000

[3] Nancy Cartwright. How the Laws of Physics Lie. Clarendon Press 1983.

[4] The 2013 Nobel Prizes. Economist p. 88 October 12th 2013

[5] Marjorie C. Malley. Radioactivity OUP 2011. Good book providing fascinating details of the development of the theory of radioactivity.

Chapter 6. Chaos and Prediction: Reality comes knocking but is ignored

[1] James Gleick. Chaos. Heinemann: London 1988

[2] R. Malhotra et al. Chaos and Stability in the Solar System. Proceedings of the National Academy of Science of the USA 98;22:12342-12343).

[3] http://www.freakonomics.com/2008/04/21/how-valid-are-tv-weather-forecasts/

[4] H. Resit Akçakaya http://www.ramas.com/mistakes.htm

[5] Nate Silver. The Signal and the Noise: Why So Many Predictions Fail--but Some Don't. Penguin Press.

CH 7 The Human Brain and DNA: the limits of Reductionism

[1] David Taffel. Nietzsche Unbound. Paragon House 2003

[2] John Horgan. The End of Science: Facing the Limits of Knowledge in the Twilight of the Scientific Age, Broadway Books 1996.

[3] Charles Choi
http://www.scientificamerican.com/article/fact-or-fiction-cockroach-can-live-without-head/ March 15th 2007

[4] Charles Darwin. The Descent of Man, and Selection in Relation to Sex. Published 1871

[5] Richard Dawkins. The Blind Watchmaker. Norton and Company, Inc. 1986

[6] Jonathan Keats. Thought Experiment. Wired July 2013 UK Edition p.128-135

[7] Guardian 7th July 2014. Arguments over brain simulation come to a head.

[8] Schmutz (2004). Quality assessment of the human genome sequence. Nature 429: 365–368. doi:10.1038/nature02390. PMID 15164052

[9] Washburn, M. P.; Wolters, D.; Yates, J. R. (2001). Large-scale analysis of the yeast proteome by multidimensional protein identification technology. Nature Biotechnology 19 (3): 242–247

[10] New Scientist 30th April 2016 p. 15. Schizophrenia's foundations may be laid down in the womb.

[11] Angela Herring. Wired UK Edition. The Human Proteome 12.12 p,136.

Chapter 8. The Limits of Observation

[1] Albert Einstein Geometry and Experience. Methuen, London 1922
[2] Ian O. Angell and Dionysios S. Demetis Science's First Mistake by Bloomsbury Academic 2010.

Chapter 9. Engineering: the red-headed step-child of Science (and just as unfairly judged)

[1] Eugene S. Ferguson. Engineering and the Mind's Eye. MIT Press 2001.
[2] CNBC. June 11, 2013
[3] Arthur C. Clarke in his essay: Hazards of Prophecy: The Failure of Imagination
[4] Langmead, Donald; Garnaut, Christine, eds. (2001). Encyclopedia of architectural and engineering feats. p. 254.
[5] Vannevar Bush. Endless Horizons. 1946, page 52-53
[6] Alan Colquohon, Typology and the Design Method, in Perspecta:The Yale Architectural Journal 1969;12:71-74

[7] https://failures.wikispaces.com/Hartford+Civic+Center+(Johnson))
[8] New York Times 1990, May 1 and 10th editions, June 15th, 28 and 29th editions
[9] Richard Feynman. What do you care what other people think? New York 1988 p 226-232.
[10] John Oberst. http://www.nbcnews.com/id/6872105/ns/technology_and_science-space/t/deadly-space-lessons-go-unheeded/#.Ujqjl9K-nN0 1/26/2005
[11] Edwin Layton Mirror Image Twins: The Communities of Science and Technology in 19th-Century America. Technology and Culture 1971;12:562-580
[12] Herald Tribune March 27th 1982
[13] Missile Defense: Cost Increases Call for Analysis of How Many New Patriot Missiles to Buy (Letter Report, 06/29/2000, GAO/NSIAD-00-153)
[14] http://www.gao.gov/new.items/d06356.pdf
[15] (http://www.nationaldefensemagazine.org/blog/lists/posts/post.aspx?ID=661)
[16] (http://www.nationaldefensemagazine.org/blog/lists/posts/post.aspx?ID=661)

[17] (http://afpprinceton.com/2011/04/america%E2%80%99s-vanishing-military-how-acquisition-failures-threaten-u-s-military-might/)

[18] (http://spectrum.ieee.org/static/weaponstable1)

[19] From a McKinsey and Company report by John Dowdy and John Niehaus in 2010. MoD_US_Equipment.pdf

[20] John Orton. The Story of Superconductors University Press 2004.

[21] Mario Bertoletti. The History of the Laser. Institute of Physics Press, 2005

Chapter 10. Medicine and the Body

[1] M. and H. Faclam and S. Grady. Modern Medicine publ. by Facts on File 2004.

[2] David Wooton. Bad Medicine. Oxford University Press 2006

[3] Roy Porter. Blood and Guts. Allen Lane 2002

[4] Thomas McKeown The Modern Rise of Population. Edward Arnold 1976.

[5] Wrigly E and Scofield R. The Population of England 1541-1871: A reconstruction. Cambridge University Press 1981

[6] Thomas McKeown. The Role of Medicine. Basil Blackwell Oxford 1979.

[7] Emily Grundy. The McKeown Debate: time for burial in Int. J. Epidemol 2005;34:529-573

[8] John Mann. The Elusive Magic Bullet. Oxford University Press 1999

[9] Druin Burch. Taking the Medicine Chatto and Windus 2009

[10] Steel K, Gertman PM, Crescenzi C, Anderson J. Iatrogenic illness on a general medical service at a university hospital. N. Engl. J. Med. 304 (11): 638–42) 1981

[11] Estimates of deaths per year caused by iatrogenesis in the United States alone are estimated at around a quarter of a million, see Starfield B. (July 2000). Is US health really the best in the world? JAMA 284 (4): 483–5

[12] Ben Goldacre. Bad Pharma. Fourth Estate 2013. Very readable book that deals extensively with problems in the pharmaceutical industry

[13] Turner et al. Selective Publication of Antidepressant Trials and Its Influence on Apparent Efficacy. New England Journal of Medicine 2008;358:252-260

[14] Begley and Ellis. Drug development: Raise standards for preclinical cancer research. Nature 2012;483(7391):531-533

[15] Bourgeois et al Outcome Reporting Among Drug Trials Registered

in ClinicalTrials.gov. Annals of Internal Medicine 2010;153:158-166

[16] Bero L. et al. Factors associated with findings of published trials of drug-drug comparisons: why some statins appear more efficacious than others. PLoS Med 2007;Jun 5;4(6):e184

[17] Economist 7th April 2012 p. 78

[18] Jonah Lehrer. Dead-end experiments, useless drugs, unnecessary surgery. Why science is failing us. Wired 20.01 p. 104

[19] Ibid p. 107

[20] [http://www.achievement.org/autodoc/printmember/mar1int-1]

Chapter 11. Medicine and the Mind

[1] Robert Whitaker. Anatomy of an Epidemic: Magic Bullets, Psychiatric Drugs, and the Astonishing Rise of Mental illness in America. Broadway Books 2011

[2] M. Foucault. History of Madness. Routledge 2006

[3] F. Alexander and S. Selesnick. The History of Psychiatry. George Allen and Unwin 1967

[4] Karl Menninger et al. The Vital Balance: the Life Process in Mental Health and Illness. New York Viking Press P.9 1964

[5] Thomas Szasz. The Myth of Mental Illness: Foundations of a Theory of Personal Conduct. Boston, Harper and Row 1961

[6] Hannah S. Decker. The Making of DSM-III: a diagnostic manual's conquest of American psychiatry. OUP 2013. Good book on the history and making of DSM-III though it is clear that the author's sympathies lie with Spitzer's DSM categorisers.

[7] R.D. Laing. The Politics of Experience. Pantheon Books 1967

[8] David L. Rosenahn . On Being Sane in Insane Places. Science 179:250-258 (1973 Jan 19th)

[9] P. Ash. The reliability of psychiatric diagnosis. Journal of Abnormal and Social Psychiatry 1949;44:272-276

[10] Lisa Appignanesi. Mad Bad and Sad . Virago 2008

[11] Wired 12.12 p. 138

[12] David Kupfer and Darrel Regier. Neuroscience, Clinical Evidence, and the Future of Psychiatric Classfication in DSM-5. Annual Review of Clinical Psychology 2011;168:672-674

[13] Guardian 26th August 2006.

[14] WHO 2000 Global Burden of Disease

Chapter 12: Social Sciences: Down and Dirty

[1] David Hume. An Enquiry Concerning Human Understanding. Oxford World Classics 2008

[2] Glaser, B. and Strauss, A.L. The discovery of grounded theory: Strategies for qualitative research. Aldine: Chicago 1967

[3] Kubler-Ross E. On Death and Dying. What the Dying Have to Teach Doctors, Nurses, Clergy, and Their Own Families, reprinted., Simon & Schuster: New York. 1997

[4] Holloway. I, Wheeler, S. Qualitative Research in Nursing. 2nd edn. Blackwell, Oxford 2002

Chapter 13 Long Time-Scale Controversies: Evolution and Climate Change

[1] S. Bowman. Radiocarbon Dating (Interpreting the Past). British Museum Press 1990

[2] E.O. Wilson. The Diversity of Life. Allen Lane 1992

[3] A. Hoffman. Arguments in Evolution. OUP 1989

[4] Derek Hough. Evolution. Book Guild Publishing 2007

[5] Gordon Rattray Taylor. The Great Evolution Mystery. Secker and Warburg 1983. Though a little long in the tooth now, this book at least can be read by the non-specialist who is interested in some of the scientific arguments against Neo-Darwinism being the only mechanism driving evolution. Indeed, this book is a greater compendium of many counter arguments than I have had room to go into in this chapter, though not all of them are necessarily compelling. It should be emphasised that Taylor did not subscribe to creationist or other Christian fundamentalist beliefs, as indeed neither do I.

[6] Joanna D. Haigh, Ann R. Winning, Ralf Toumi, Jerald W. Harder. An influence of solar spectral variations on radiative forcing of climate. Nature 2010; 467 (7316): 696–9.

[7] J. Houghton Global Warming: The Complete Briefing publ. Cambridge University Press 1997

[8] Economist magazine March 30th 2013 p. 81 and see also August 23rd 2014 p. 69.

Chapter 14 Risk in Science

[1] Marjorie C. Malley. Radioactivity: A History of a Mysterious Science. OUP 2011

[2] Robert Jungk. Brighter than a Thousand Suns. Penguin Books 1982 (originally published 1956)

[3] J-P. Blaizot et al. Study of Potentially Dangerous Events during Heavy-Ion Collisions at the LHC: Report of the LHC Safety Study Group CERN 2003–001

[4] http://press.web.cern.ch/backgrounders/safety-lhc

[5] http://lsag.web.cern.ch/lsag/LSAG-Report.pdf

[6] Meselson M, Guillemin J, Hugh-Jones M, et al. The Sverdlovsk Anthrax Outbreak of 1979 (November 1994) Science 266 (5188): 1202–8t

[7] Druin Burch. Taking the Medicine. Chatto and Windus 2009

[8] Rebecca Skloot. The Immortal Life of Henrieta Lacks. Pan 2010. Excellent and accessible read that details not only what has happened to Henrietta Lacks' cells since her death but also the distressing effects on her surviving family of this exploitation.

Chapter 15. So Now What?

[1] A Companion to Nietzsche, edited by Keith Ansell Pearson, Blackwell 2006

[2] as quoted by A. Willis 2009, 'Immortality only 20 years away'. Daily Telegraph 22nd September 2009

[3] Rupert Sheldrake. The Science Delusion. Coronet 2012

[4] Wired (05.13 p. 101)

[5] Sunday Times 6th October 2013 p.3

Appendix: Some Fundamental Mysteries Still not Explained by Science

[1] National Geographic Magazine May 2010. The Secrets of Sleep

[2] Crick F. Memory and Molecular Turnover Nature 312:101 1984

[3] R. Lewin. Is Your Brain Necessary. Science 1980;210:1232

Acknowledgements

Firstly, I would like to thank my many friends and colleagues who have put up with my musings and occasional rants on this subject over the last few years. In particular I would like to thank the following for reviewing the manuscript and making suggestions for its improvement: Prof. David Wyper, Don Ross, Dr. Marcel Strauss, Dr. William Holmes, Ann McKinlay, Dr. Roderick Duncan, Dr. Jim Patterson, Dr. Doug Small, Dave Sutton, Graham McKinlay and Stewart Horne. They were all encouraging of the project but, needless to say, they did not necessarily agree with all of the contents all of the time. Any remaining errors are all my own.

I am also deeply grateful to Elizabeth, Ann, Juanita and Aurora for giving me the time and space to think my heretical thoughts.

About the Author

Barrie Condon has degrees in Physics, Oceanography and Nuclear Medicine. Now retired, he spent over thirty years as a Medical Physicist working for the UK's National Health Service as a Consultant Scientist. He also held an Honorary Professorship with the University of Glasgow, and at the peak of his career was responsible for the work of over 300 scientists and technologists. His own research work has mostly been on the neuroscience applications of medical imaging techniques, including MRI and Nuclear Medicine.

He is an author on over 80 full research publications in peer reviewed journals such as the Lancet and the British Medical Journal and on over 200 national and international conference presentations.

scienceforheretics@gmail.com

Further reading

Sparsile Fiction

The Promise
When promises can mean lives

L. M. Affrossman

Eight-year-old Daniel Gallagher is dying. His mother refuses to believe it, and Daniel is too afraid to make her listen. Then, by chance, he meets Sadie Gordon, an eighty-one-year-old survivor of Stalinist Russia, and everything changes.
It isn't long before he has captured her heart and she can't imagine life without him. Only then does the secret that has made Daniel old beyond his years come tumbling out.
And, in a moment of confusion, he exacts from her a promise to accompany him to heaven when he dies. Sadie is left with a terrible choice. Should she compromise the moral beliefs of a lifetime or betray the trust of a dying child?

Sparsile Fiction

Simon's Wife
A secret history

L. M. Affrossman

Three decades have passed since the death of Jesus of Nazareth on a
Roman cross, and Judea is teetering on the brink of apocalypse. Caught up in
the horror that will inspire the Book of Revelation, a young woman is fighting
for survival.

In the dark days that follow the Roman devastation of Jerusalem in AD 70,
nineteen-year-old Shelamzion bat Judah finds herself captured and awaiting
both her own execution and that of her husband, former rebel-leader Simon
bar Gioras. Alone and forgotten, there seems little reason to go on living, yet
a strange friendship begins to grow between Shelamzion and her austere, old
Roman jailor, Fabius Cornelius Grammaticus.

With his pretensions to be recognized as an historian in the style of Livy, it
is to her he turns to record the true version of events behind the insurrection
in Judea that led to the destruction of her country. Time is running out, however, and unknowingly history is being rewritten by a traitor's hand.

Comics and Columbine
An outcast look at comics, bigotry and school shootings

Tom Campbel

THE SCHOOL SHOOTER WHO DIDN'T SHOOT.

Growing up an autistic loner Thomas Campbell's schooldays were a living nightmare of bullying and abuse that saw him in psychiatric care by age 8.

The target of entire classrooms, he developed a lifelong hatred of all things educational. This hatred – the shared thinking of the school shooter – has gifted him with a unique insight into the slaughter we are witnessing in our schools now.

For the first time a book is written from the perspective of the classroom avenger, one that explores their distorted thinking and reveals the 'socially acceptable' evils that provoke such a lethal response.

Lightning Source UK Ltd.
Milton Keynes UK
UKHW011041170121
377170UK00001B/41